Hands-On Environmentalism

Hands-On
Environmentalism

Brent M. Haglund ▪ Thomas W. Still

ENCOUNTER BOOKS
SAN FRANCISCO

First paperback edition published in 2005 by Encounter Books, an activity of Encounter for Culture and Education, Inc., a nonprofit, tax exempt corporation.

Encounter Books website address: www.encounterbooks.com

Manufactured in the United States and printed on acid-free paper.

The paper used in this publication meets the minimum requirements of ANSI/NISO Z39.48-1992 (R 1997) (*Permanence of Paper*).

Library of Congress Cataloging-in-Publication Data

Haglund, Brent M. and Still, Thomas W.
    Hands-on environmentalism / Brent M. Haglund and Thomas W. Still.
        p. cm.
        Includes index.
        ISBN 1-59403-085-5 (alk. paper)
Environmental responsibility—United States. 2. Environmental policy—United States. 3. Environmentalism—United States. I. Title.

GE195.7 .H34 2005
333.72 –dc 22
2005050057

10 9 8 7 6 5 4 3 2 1

Contents

1

The Environmental Nanny

The top-down, angst-ridden environmentalism of the late twentieth century puts power in the hands of bureaucrats who discourage innovation and punish people, businesses and communities for trying to make improvements. What's needed is an environmentalism that harnesses the power of citizens. And we need it fast.

"WHEN WE TRY TO PICK OUT ANYTHING BY ITSELF," legendary conservationist John Muir observed a century ago, "we find it hitched to everything else in the universe." Muir was right. Our complex natural world is a maze of seen and unseen connections, constantly evolving and adapting, forever compelling its myriad physical and biotic elements to work together.

Sadly, the prevailing system for safeguarding the natural world today—a command-and-control system—is the antithesis of how that world actually works. The bulk of environmental law and regulatory process in the early twenty-first century is linear, segregated and inflexible. Muir's theory that everything in nature is connected to everything else doesn't seem to apply to conventional regulatory thinking about how man is linked, for better and worse, to the environment. The people who created today's regimented structure of environmental protection are out of touch with the primary truth of the natural world they hope to serve.

Too much of environmentalism has become the province of "experts"—in government, business and the leading enviro-political groups—rather than citizens. We have created a cadre of professionals, including bureaucrats and "enviro-pols," who have usurped from ordinary people the job of managing the environment. Those experts may have noble intentions, but they aren't doing a good job of running things.

The political environmentalism of the past thirty to forty years was born of necessity. Business-as-usual was not protecting the air, water and land; there were grievous examples of pollution crossing local and state borders, which invited action by Congress and federal regulators. The Clean Air, Clean Water and Waste Management Acts and the creation of the Environmental Protection Agency were among Washington's responses.

The regulatory actions of the 1960s and 1970s were welcomed as medicine to help cure a throwaway society. Even in a nation blessed with natural resources, it was time to stop throwing away and start conserving. The natural human tendency to overdo a good thing—if one pill works, let's take four; if a gallon of fertilizer helps, let's dump ten—had reached an illogical limit. Alarmed by dirty water and foul air, people realized that the time was right for environmental activism. The first Earth Day on April 22, 1970, the brainchild of Gaylord Nelson, a Democrat U.S. senator and conservationist from Wisconsin, was a public expression of the belief that citizens should be at the heart of bringing about a cleaner world.

Over the decades, the cure became something of a disease itself. What began as a check on environmental abuses grew into a command-and-control system that inhibited innovation and technological progress while widening the gulf between people and the natural world—a world that in fact includes people. The state became an environmental nanny, constantly wagging a scolding finger but rarely encouraging anyone to do better or teaching them how.

The regulatory system that exists today is disconnected not only from nature but also from other branches of human activity. In art, music, science, commerce, sports and just about any other human endeavor, the goal is continuous improvement. In nanny environmentalism, by contrast, the goal is compliance with minimum standards rather than the achievement of measurable gains. The tools are punitive rather than incentive-based. Bullying is part of the process. Partnership is viewed as a race to the bottom rather than a way to lift everyone's performance. Success is measured like a traffic cop working under a ticket quota: How many fines have we levied? How many noncompliance notices have we mailed out?

Break down the term "command-and-control" and here's what you get:

- "Command"—Lawmakers and bureaucrats do the commanding. The commands go like this: "Do this and do it now, or else." Or, "Stop doing that now, or else."
- "Control"—Bureaucrats, often at the urging of enviro-political groups, do the controlling. Control is the "or else" part. "Pay this fine." "Forget about your project; it's dead." The role of the landowner, the company, the village is to obey. Or else.

Decades of command-and-control have produced these results:

- Citizens who want to do something about the environment have been muscled out of the picture. The bureaucratic controllers, jealous of their prerogatives and infused with the idea that their way to do something is the only way, have systematically frozen out all other potential players. (Chapter 7 provides detail on this.)
- Companies that want to launch environmental projects have to dance to the tune of the bureaucratic piper. That has been a serious disincentive.
- Federal or state enviro-goons jerk around municipalities that want to map their own plans.

As a consequence, individual citizens have become discouraged. Companies have, for the most part, retreated to an entirely defensive and risk-averse position. They spend a lot of money and energy warding off or trying to soften the commands from the controllers, and they rarely stick out their necks to try anything new that might improve the environment. Municipalities have, in many cases, given up.

Command-and-control has turned officialdom into adversaries of the people and organizations who are vital to achieving environmental results. Adversaries cannot get much done together. A command-and-control regulatory system does not inspire people to do better. Instead, it teaches them to expect that when they make mistakes, even in an effort to improve, their hands will be slapped and they will be sent to bed without supper.

Command-and-control or "nanny" environmentalism has evolved into a regulatory system built upon mistrust. Government doesn't trust the people. Business doesn't trust the government. The "Big Green" environmental groups don't trust anyone. This mistrust creates a sense of public detachment, even cynicism, about environmental protection. It makes people more susceptible to alarmism. It inhibits the human spirit and sometimes makes us feel bad about ourselves and our potential, rather than encouraging us to think of humanity as a joyous and rational part of the interconnected natural world that Muir described.

Command-and-control environmentalism has become a victim of its own early successes. Partly as a result of federal and state regulations and enforcement, effluents no longer pour from industrial stacks, chimneys and waste pipes. Garbage is no longer burned openly. Rivers don't catch fire these days. The dumping of untreated sewage or wastewater is rare, rather than the rule. Reforestation has become common practice for timber companies, to the point that there may be more wooded and forested acreage in America today than there was in colonial times. Air pollutant levels are lower, even in major cities.

Some of this was accomplished not because of command-and-control environmentalism, but in spite of it. Communities and companies resolved to clean up as they learned more about the effects of their polluting behavior on the environment, and about how continuing that behavior would harm future generations. The guiding and sometimes heavy hand of government deserves some of the credit, but so does Adam Smith's "invisible hand."

Now, early in the twenty-first century, many people are asking: How can we move to the next level of environmental stewardship? The command-and-control model isn't designed to get us there. Much of what remains to be done lies outside the reach of government regulators or cannot be accomplished by a program imposed from on high.

Consider the record of some major environmental programs, created with the best of intentions, but now mired in bureaucracy, litigation or both. Examples include Superfund, the Endangered Species Act, the Clean Water Act, and the politics surrounding global warming.

• Superfund, the federal program to clean up hazardous waste sites, was envisioned as swift and short-lived—a concentrated burst of environmental disinfectant. Congress authorized the EPA to clean up waste sites immediately and then bill the responsible parties. "It was supposed to deal rapidly with emergencies by cutting through red tape," wrote Richard L. Stroup of the Montana-based PERC (Political Economy Research Council), a center for free-market environmentalism. "But it failed." Instead of cutting red tape, Superfund has created green tape. Congress allowed the EPA, through Superfund, to override common-law concepts of property rights with bureaucratic control. As a result, the program became a paradise for lawyers and a hellish trap for true environmentalists. Superfund cleanups average twelve years and $30 million, a good part of it spent on legal fees. Superfund projects can also be divisive, ripping communities apart rather than bringing them together.

The Hudson River dredging project is a case in point. No one disputes that a forty-mile stretch of the Hudson River north of Albany, New York, was polluted by polychlorinated biphenyls (PCBs) over the years by two General Electric plants that discharged them—legally at the time—into the water. The last of 1.3 million pounds of PCBs was discharged in 1977, when the federal government banned the cancer-causing chemical. Sharply disputed for more than two decades, however, is how to clean the Hudson.

The Environmental Protection Agency intends to dredge the polluted part of the river at a cost of about $500 million. The EPA confirmed this decision in late 2001 and kept plodding along despite evidence that the river is cleansing itself and that remaining PCBs are hidden under layers of silt. Some citizens of the Hudson River valley support dredging, but others vehemently oppose it. The political standoff created by the EPA's uncompromising approach over more than twenty years has ended any chance of a locally negotiated solution and has diverted citizen energy and resources into lawsuits and a war of press releases, rather than a search for ways to work together.

Citizens who oppose dredging of the Hudson say it will stir up dormant PCBs, ruin resurgent sport fishing, harm local businesses that have already suffered from years of uncertainty, create 2.65 million cubic yards of sludge with no place to put it, destroy 97 acres of wetlands and generally shut down the river for years. An organization called Citizen Environmentalists Against Sludge Encapsulation (CEASE) expects the fight to drag on in the courts.

"There's a lot more fear about what's going to happen to our town during dredging than if the PCBs stay where they are," said Sharon Ruggi, deputy supervisor of Fort Edward, New York. "We think dredging will create far more environmental damage than it will solve. But what we have here is a political fight that has nothing to do with the science and never did. Common sense has no place here."

Ruggi, a member of CEASE (www.nodredging.com), says that not even public meetings packed with citizens who overwhelmingly oppose dredging have been able to sway the EPA from its course. "For people who have never dealt with the EPA, [the agency] is kind of viewed as the guys with the white hats. But now that we've dealt with them, I say they're more like the uninvited guest who comes and never leaves."

The Hudson River dredging would be the largest in the nation's history, but an even larger project has been proposed in northeast Wisconsin's Fox River valley, where the culprit is PCBs discharged legally by paper factories from the 1950s through the early 1970s.

• The Endangered Species Act is a case study in perverse incentives. Property owners across the United States now manage their land with the express intent of keeping endangered or threatened species out because they fear losing their rights if some such species is discovered there. The act's track record is poor: Fewer than a dozen of more than 1,400 listed species have recovered since the federal law was passed in 1973. The act must be reformed so that landowners can become partners in saving endangered species, not treated as enemies without rights.

The Endangered Species Act is counterproductive for animals as well as people. There may be no better example than the federal shutoff of irrigation water to farmers around Upper Klamath Lake along the Oregon/California border, where a misguided effort to save endangered suckerfish and salmon devastated farmers—and may have hurt the fish also.

In April 2001, the National Marine Fisheries Service and the

U.S. Fish and Wildlife Service ordered irrigation water cut off to 1,400 farmers because the agencies believed that water levels in the lake must remain high in order to dilute the runoff of agricultural and other sediments. A higher water level would give the suckerfish and salmon a better chance to prosper, the agencies claimed.

The prosperity of the farmers was not a concern. The loss of water cost the local economy 2,000 jobs and $130 million, according to a joint report by the University of California at Berkeley and Oregon State University. Many of the farmers who suffered a loss were descendants of World War I veterans who had obtained and worked the land at the inducement of the federal government. These veterans were promised title to the land and perpetual water rights if they would farm it. The farmers kept up their end of the bargain, turning the Upper Klamath valley into a green oasis. The feds did not. People whose families had been stewards of the land for three or four generations were forced to sell or even abandon their property after the water was shut off.

Then came the bitter irony: the water shutoff probably didn't help the fish. In a January 2001 report, the National Academy of Sciences said there was "no substantial scientific foundation" for claims that the Upper Klamath water level needed to be kept high. In fact, raising it probably harmed fish by simultaneously raising water temperatures, especially around cold streams where fish go to avoid the summer heat.

The federal Bureau of Reclamation announced about a month later that it would release irrigation water from the lake to area farmers in time for the 2002 growing season. But it was too late to save the farmers who had lost their land to environmental zealotry.

"We are ecstatic about the [NAS] study, but the problem I have is that I lost a lot of good neighbors and friends, including my next-door neighbor who lost his farm of thirty years," said Bob Gasser, who lives in the Upper Klamath area. "Incomplete science took his livelihood, his home and his children's education."

There's still a chance for a belated happy ending to this story. Interior secretary Gale Norton organized a working group to find a lasting and balanced solution. This resulted in a partnership with the Klamath Basin Rangeland Trust, organized by Oregon ranchers Jim Root and Kurt Thomas, which will save water through voluntary participation by farmers. The nonprofit trust will also implement conservation and wetlands restoration measures on lands no longer irrigated in order to improve water quality.

• Threatened by lawsuits filed mostly by environmental groups in more than thirty states, the EPA has belatedly decided to enforce a little-known provision of the 1972 Clean Water Act that requires cleanup plans called "Total Maximum Daily Loads" for many rivers, lakes, streams and watersheds. Congress and the White House have put even this tentative effort on hold—for good reasons. The TMDL is a complicated calculation of how much pollution a given body of water can receive on a given day and still meet water quality standards. To some, mainly those who have filed suits, it's high time the EPA cracked down, and they are frustrated with delays.. To others, TMDLs are the ultimate command-and-control act by a federal agency that has lost touch with economic reality, federalism, the concept of local control and the laws of nature itself. Nationwide, states must develop more than 40,000 TMDLs, a job that will cost them about $1 billion a year over the next fifteen years—and that's just the cost of writing the plans and establishing a baseline, not implementing the plans. (Paul Portney, president of Resources for the Future, and others concerned about the gaps between the public's expectations and the actual improvements in water quality point out that we in the United States are now spending billions without being able to inform ourselves reliably as to what water quality status is now or how it has changed.) Fiscally pressed states say that TMDLs violate their rights and responsibilities. Farmers and other property owners say that the EPA is running roughshod over their rights. Cities say that TMDLs will encourage sprawl by making it harder to build in existing urban areas. Some ecologists believe that TMDLs won't work because of the complex nature of water bodies, which are subject to unpredictable natural events that influence pollutant levels and the ability of the water body to cleanse itself.

In short, the EPA has taken 40,000 local challenges that could be solved by states, communities and landowners working together, and has turned them into a giant federal problem that most likely will never be solved.

• The fight against global warming has been costly, with little to show for it. That is a major conclusion reached by two authors in "The Death of Environmentalism," a 2004 paper that has caused many environmentalists to question their reliance on the political tactics of the past. Authors Michael Shellenberger and Ted Nordhaus don't downplay the effects of global climate change—in fact, they contend that it's an enormous crisis—but they are sharply critical of environmental leaders clinging to command-and-control political strategies that fail to inspire either policymakers or the public at large.

"By failing to question their most basic assumptions about the problem and the solution, environmentalists are like generals fighting the last war—in particular the war they fought and won for basic environmental protections more than thirty years ago," Shellenberger and Nordhaus write. "It was then that the community's strategy became defined around using science to define the problem as 'environmental' and crafting technical policy proposals as solutions."

The authors argue that citizens won't respond to admonitions to drive hybrid cars and use fluorescent light bulbs, but will respond to a "big vision and a core set of values" that inspire them to act on their own rather than simply being told what to do.

The conservative movement in America has succeeded, Shellenberger and Nordhaus say, because it is clear on vision and values. For better or worse, people have responded. In contrast, "environmental groups have spent the last 40 years defining themselves against conservative values like cost-benefit accounting, smaller government, fewer regulations, and free trade, without ever articulating a coherent morality we can call our own."

By framing ideas around core American values, they suggest, environmentalists can move beyond "special interest" status and chart a new course in which stewarding the land, water and air is seen as part of the political fabric, rather than as an attempt to tear it apart.

THERE IS A BETTER WAY. In some states and communities, a concept called "civic environmentalism" is taking root. It rests on the belief that the current win/lose system of retribution and risk avoidance must be replaced by a system that seeks more winners and fewer losers, less punishment for past sins and more reward for solving problems, and the replication of those "random acts of goodness" that so often spring from public-private partnerships and individuals working on their own.

Jeff Smoller, a top assistant to the secretary of Wisconsin's Department of Natural Resources (DNR), described civic environmentalism as "the freedom to try and the duty to learn." Smoller commented: "So often, our environmental arguments are about the past and retribution and not the future and new possibilities. That so permeates the conversation that we're unable to make assessments about risks, to apply incentives and to make reasonable tradeoffs."

In Wisconsin, where John Muir spent much of his youth and where Aldo Leopold wrote *A Sand County Almanac,* the process of reforming and even replacing the command-and-control structure is under way. It is a process that other states will follow with interest, given Wisconsin's history of environmental activism and its reputation for being an aggressive practitioner in the command-and-control school.

Wisconsin's "Green Tier" program is a prime example of the new way of thinking. It's an attempt to move beyond unilateral enforcement of minimum standards to partnerships that aspire to maximum levels of environmental protection. Green Tier offers a way to go beyond compliance, to address unregulated environmental problems and to restore natural resources in return for incentives that are tied to superior performance.

Green Tier is designed to focus environmental work in communities and industries, not bureaucrats, and to provide legal standing for that work. Under Green Tier, all organizations and sectors may enter into legally binding contracts that address multiple environmental goals. Green Tier uses three tools.

1. *Environmental charters.* Charters are legal instruments that define the scope of responsibility, activities, authorities and services needed to achieve superior environmental performance. They may be organized around land areas, watersheds, air quality zones, forests, political subdivisions, trade or business sectors, products, occupations, supply chains, emissions categories, species, biological concepts and more. Negotiated by private players and the state,

charters are flexible legal instruments that give standing to a party or parties to get things done.

2. *Environmental contracts.* These are enforceable contracts entered into by the state but often initiated by private parties that specify each party's commitment to superior performance. In some cases, the state or others might agree to incentives or support that is proportional to the contract's goals. Under Green Tier, the contract is the enabling and committing legal instrument used to trigger rewards for achievements or sanctions for shortfalls.

3. *Environmental management systems.* These are business systems focused on achieving environmental results. They organize procedures and resources, monitor data, adapt plans and continuously improve results. It's an evaluation system.

Incentives under Wisconsin's embryonic Green Tier system include regulatory flexibility, streamlining, technical assistance, one-stop shopping for groups that need to deal with the DNR, recognition and use of Green Tier or Green Star logos for public relations and marketing purposes. For companies, it can make environmental performance pay in a tangible way. It can align economic performance—always a corporate goal—with environmental and social performance, two standards that may define twenty-first-century capitalism.

In addition to Green Tier, Wisconsin's DNR is working with citizen groups, examining third-party environmental evaluation and facilitation, leading a multi-agency program to test and develop environmentally sustainable farming practices (the Wisconsin Agricultural Stewardship Initiative), and pursuing partnerships with business, government and environmental groups in Germany and the Netherlands. Wisconsin's DNR has also taken a lead in the Multi-State Working Group, which worked with Harvard University to host a 2003 conference in Washington, D.C., on the future of environmental regulation. The international relevance, and the imperative for environmentally conscious businesses to lead the way, was manifest in a University of Wisconsin Green Tier Seminar held in January 2004. Business and government leaders from Germany, the Netherlands and other nations made the case forcefully and clearly for moving from command-and-control to contracting for environmental performance.

Not so long ago, many in Wisconsin were convinced that the

initials "DNR" stood for "Darn Near Russia." Five miles outside any city in the state, hatred of the DNR burned white hot. To landowners all over the state, DNR staffers were horned bogeymen. A farmer would ask, "Can I build a buffer on this portion of my land instead of over there?" The answer: "No!"

A DNR field man once gave an acquaintance a nice-looking cap sporting the letters "DNR." The field man said, "Here. Wear it...if you're man enough." The few times when the guy wore it nobody jumped him, put a gunny-sack over his head and pounded on him, but he did get nasty comments and glares.

Insiders say the DNR still has more old-school disciples than new. Even so, the more inclusive and civic efforts of the Wisconsin DNR hold promise for a dramatic change. Unfortunately, the new attitude is still very much the exception in environmental thinking.

The transformation from command-and-control or nanny environmentalism to a more civic model will succeed only if people insist that it happen. They must demand that their government stop treating them like dimwits who are incapable of solving problems close to home. And they must prod their fellow citizens and their private institutions to recognize that there are powerful economic and moral arguments for preserving our complex natural environment.

"The biosphere promotes the long-term material prosperity and health of the human race to a degree that is almost incalculable," wrote Edward O. Wilson in a 2002 essay, "What Is Nature Worth?" Like Muir, he argued that it makes no sense to calculate the value of a single organism because human beings cannot fathom all the ways it may be connected to others—including man himself. "A conservation ethic is that which aims to pass on to future generations the best part of the nonhuman world. To know this world is to gain a proprietary attachment to it," Wilson wrote.

Another great conservationist, Aldo Leopold, argued nearly sixty years ago for a conservation ethic based on the same sense of personal, proprietary attachment to the larger biotic community. Today, that ethic might well be described as "hands-on" environmentalism. It is an ethic for an exciting new age.

2

"Hands-On" Environmentalism

How participatory democracy is saving the land

*Driven by a desire to re-engage in public life and frustrated with institutional-
ized "command-and-control" environmental programs, citizens are embracing
a more local, "hands-on" approach to environmental activism.*

THERE IS A MYTH THAT AMERICANS DON'T CARE. Rich Harwood
shattered that myth in his 1991 report for the Kettering Foundation, *Citizens and Politics: A View from Main Street America*. After
talking with citizens across the United States, the nationally
renowned researcher concluded that Americans aren't apathetic—
but they can be cynical, and they feel impotent with regard to
politics and public affairs. The Main Street citizens who sat down
with Harwood told him they had been pushed out of public life
and left with little room to understand, get involved and make a
difference in their community or state.

"People have gotten so disappointed that they don't want to
get involved anymore," said a Seattle woman.

"There should be a whole array of ways for people to get
involved—and there just aren't," added a woman in Dallas.

His conversations led Harwood to observe that "citizens
engage in specific areas of public life when they believe they can
make a difference." In fact, many may choose to sit it out unless
"they believe a political compact exists that suggests: 'When I participate there will be at least the possibility to bring about and
witness change.' By and large, citizens do not believe this compact
is present in most areas relating to political action today."

Nearly fifteen years after *Citizens and Politics* called for a reconnection of Americans to public life, the nation is still crafting its compact for participatory environmentalism. To Harwood, the struggle to create a new and more "hands-on" environmental ethic is tied to core feelings about stewardship as well as the basic American value of individual responsibility. Public concerns about the polluted *civic* environment have finally intersected with public worries about the *physical* environment.

"We have come to rely so much on rules and regulations that we're squeezing human nature out of the equation," Harwood said. "It has put us in a box; we have come to believe we can no longer operate on our own. We have forgotten how to talk to one another. So we resort to blunt instruments such as laws and lawsuits to control ourselves."

Those who question environmental rules or procedures are quickly turned into "cartoon figures," Harwood continued. "The rancher is portrayed as a self-interested, capitalist heathen who only wants to rape the land, when, in reality, most ranchers care very much about protecting the land. The chemical plant manager is portrayed as inhuman, even if he only wants to run his plant safely and profitably." Conversely, Harwood said, people who have legitimate concerns about suburban growth patterns or water quality or other environmental problems are quickly caricatured as neo-hippies, "throwbacks to the 1960s, only with shorter hair."

And yet, Harwood said, a path through the civic jungle is being cleared. Increasingly, citizens are rejecting environmental stereotypes and rethinking tired models of behavior.

"People constantly talk about the fact that we can't legislate responsibility. You can only legislate the punishment for not taking responsibility," he said. "Similarly, you can't legislate good judgment or affection for the environment. But you can provide incentives. There are common interests and a common purpose, but don't give up on self-interest as a powerful motivator."

Like the people who were interviewed for *Citizens and Politics* in 1991, the typical American today wants to be a more active environmental citizen—if only Big Government, Big Business and the Big Green Lobby give him or her a chance to engage.

Fortunately, some policymakers have taken notice and want to remove barriers to true public engagement.

• "Protecting our nation's wild places and endangered species and precious resources depends on our abilities to develop partnerships with private landowners and groups," interior secretary Gale Norton said on April 12, 2001, at a "Private Conservation Day" celebration in Washington, D.C. "We must involve them in our decision-making process. We should take advantage of their ability to test innovative alternative approaches to conservation."

• "For too long, environmental policy has focused only on restrictions, regulations and direct government ownership," said Fred Smith, president of the Washington-based Competitive Enterprise Institute and cofounder of the Center for Private Conservation. "We've focused on what private people sometimes do wrong on our planet and not enough on how individuals and groups have done things right."

• "Conservatives need to acknowledge the public's need for progress, while liberals must accept the fact that government needs new and more versatile tools to solve today's changed array of environmental problems," said Debra S. Knopman, director of the Progressive Policy Institute's Center for Innovation and the Environment. "First generation regulation…is too slow and inflexible to capture technological innovation in quick-moving markets. It is also too prescriptive to engage landowners and deal with small, diffuse sources of pollution and too narrow to mesh well with land use, energy, transportation and agricultural policies."

• "There's a hunger for connection," said Heather Mann, executive director of the Urban Open Space Foundation in Madison, Wisconsin. "It's born in part out of a frustration with the status quo, but also out of a commitment to do something more for the environment in their community. People want to do more, and will, if you give them an opportunity."

PUBLIC OPINION POLLS ON THE ENVIRONMENT capture a sense of that "hunger." Invariably, people who are asked if a clean environment matters to them or should matter to their neighbors will say "yes." But the answers get trickier when those same people are asked about how much they're actually involved in cleaning up the environment—or if the right people are making the decisions.

In April 2000, the Gallup Poll asked: "Which of the following should have primary responsibility for solving our nation's environmental problems: The government or business and industry or citizens' groups and individual citizens." It was a statistical dead heat: 34 percent picked the government; 33 percent said business and industry; 32 percent said citizens' groups and individuals. Eight years earlier, the same Gallup question yielded somewhat different results: 44 percent wanted the government to clean things up; 20 percent wanted business to do it; 27 percent picked citizens' groups and individuals (www.PollingReport.com). These numbers indicate an erosion of public confidence in the "government knows best" approach to solving environmental problems.

• Gallup, CNN and *USA Today* have asked this question in five nationwide surveys since 1989: "Do you consider yourself an environmentalist, or not?" The "yes" responses had slid from 76 percent to 50 percent and the "no" responses had climbed from 20 percent to 48 percent by 1999 (www.PollingReport.com). Perhaps the same people who say they want to do more to help the environment don't know how to get meaningfully engaged in ways beyond recycling their trash or riding in a car pool. Some people may feel they're not worthy of the title because they don't do enough, while others may reject the label because it comes with baggage, deserved or otherwise.

• A January 1999 poll by Yankelovich Partners for CNN and *Time* magazine tested how Americans feel about the balance between regulation and private property rights (www.PollingReport.com). "If you had to choose, which is more important: the ability of individuals to do what they want with the land they own, or the ability of government to regulate residential and commercial development for the common good?" More than two-thirds (69 percent) chose individual rights while 25 percent picked regulation, with the rest undecided. Results for people living in rural America (73 percent for individual choice) were only slightly higher than for those who live in cities (69 percent) and suburbs (65 percent). Decades of government regulation have failed to convince Americans that a federal program is better for the land than a deed in the county courthouse.

• In a nationwide poll in November 1997, the Pew Research Center for the People and the Press asked respondents how often they car-pool, recycle newspapers, aluminum or glass, adjust the temperature in their house to save energy, shop for recycled paper or plastic products, or buy organic or pesticide-free foods. About 83 percent said they regularly or sometimes recycle, 78 percent tweak their home thermostats and 69 percent shop for recycled products. However, only 30 percent regularly or sometimes ride in car pools and 47 percent shop for organic or pesticide-free foods. People take part in environmental activities that are easily accomplished or convenient, but they won't be forced into activities that don't make sense for them and their lifestyles.

Americans want to do what's right by the environment, but they're not exactly sure how, and they're not convinced that command-and-control environmentalism is the answer. It's not that they don't care, as Rich Harwood demonstrated in *Citizens and Politics* fourteen years ago, but that they're unhappy with the status quo.

Lynn Scarlett, who is currently assistant secretary for policy, management and budget at the U.S. Department of the Interior, argues that Americans want a "new environmentalism" to replace an old approach that was punitive, prescriptive, process-focused and dominated by partitioned decision making. "What we want is an environmentalism that focuses on performance, progress and results; enhances cooperation rather than conflict; and offers more holistic decisions that give expression to the many values people hold," Scarlett said. "A new environmentalism…is a set of institutional arrangements and opportunities that tap local knowledge, foster tailored creativity and innovation, inspire folks to pursue environmental values, create a context for cooperation, and provide decision settings that foster a holistic look at problems, values and opportunities."

Across America, that kind of new environmentalism is bubbling up in the often-overlooked activities of thousands of citizen groups and millions of citizens. These citizens are building a participatory environmental movement that emphasizes community partnerships going beyond segmented "stakeholder" conversations. It's not free-market environmentalism, which depends entirely on the whims of the individual and the economy, but it's certainly not command-and-control environmentalism, which seeks to subjugate property rights to the will of the state.

It's called "hands on" environmentalism, and it's working.

Because it's based on values such as local control, personal responsibility, government accountability and economic opportunity, this new vision of environmental activism is challenging the command-and-control model that has dominated thought about environmental correctness for the past three decades. The command-and-control model is built upon a belief that only top-down regulation by an omniscient central bureaucracy can prevent ill-informed, selfish or rapacious people from fouling their own air, water and land. It draws on the theories of eighteenth-century economists such as Thomas Malthus to perpetuate a belief, unfortunately shared by too many Americans, that there are static, unconquerable limits to the earth's resources and to the human ability to rejuvenate our world.

Command-and-control environmentalism has relied on "Chicken Little" predictions about falling skies and assorted "crises-of-the-month" to capture the attention of the news media and the public—and to raise money for its causes. It has too often substituted political process for sound science. It has drifted away from the admirable ideals of public participation that motivated the nation's landmark environmental laws, turning them (and the agencies that enforce them) into platforms for lawsuits, social and economic conflict, and one-size-fits-all bureaucracy. Finally, it has turned its back on the notion that economic prosperity is essential for environmental stewardship.

The emerging alternative to the political environmentalism of the late twentieth century is rooted instead in the belief that the core environmental responsibilities and rights must rest not with distant governmental bureaucracies and philosophical zealots, but rather with the local citizens and stewards closest to the resources. This participatory environmentalism holds that there are limits to what governments and bureaucracies can and should do. Its intellectual roots rest in the rich and enduring land ethics of Aldo Leopold, the great Wisconsin conservationist and author of *A Sand County Almanac*.

At the most basic level, Leopold believed that people—acting as individuals or collectively, with proper incentives and a willingness to test their results—could do more good for their natural surroundings than rows of statute books, stacks of legal briefs or roomfuls of government bureaucrats. He provided a simple but comprehensive guide to what constituted real environmental responsibility. "Examine each question in terms of what is ethically and esthetically right, as well as what is economically expedient," he advised. "A thing is right when it tends to preserve the integrity, stability and beauty of the biotic community. It is wrong when it tends otherwise."

Today, those advocating the broader, more local and hands-on environmentalism have embraced Leopold's conviction that true environmentalism must be practiced and can best be sustained at that point where people and the land come into contact.

The Civic Environmentalism Working Group notes: "People live in real places, not some abstraction called 'the environment.'" Arguing that "love of place is an important source of civic attachment and civic commitment," members of the CEWG believe "the people who live in a particular place should, to the extent possible, make the crucial decisions about common issues involving its physical resources and public space." They conclude: "By doing so, they develop their capacity to deliberate about the subtle and difficult choices such decisions necessarily involve. The character and quality of the citizenry is improved by means of its effort to improve its physical surroundings."

Participatory environmentalism seeks to give citizens "real authority to make real decisions to do real things," in the words of Jeffrey Salmon, founder of the Civic Environmentalism Working Group. Granted, that exercise of "real authority" may not make everyone happy—especially established interest groups. As Salmon points out, "Resistance to local autonomy or to local solutions (however effective these may be for the environment) may come not only from federal government agencies but from national environmental groups with an established national agenda."

The old environmentalism has contributed to the loss of local control, not only for citizens but for local governments that often find themselves helplessly aligned against the Environmental Protection Agency, the Army Corps of Engineers or a state environmental department. Such policymaking also tends to put politics and process ahead of science, and to view citizens as part of the problem rather than a source of the solution.

Participatory environmentalism tries to create space for what

Salmon describes as a core function of democracy: the messy debates between diverse interests over the best possible solution to a given problem. The results can contradict what the command-and-control system might envision, or run counter to what individual property owners, acting alone, might believe is in their best interests. The process does, however, produce an outcome that is more democratic and thus more sustainable.

Consider this description of "civic environmentalism" by Carmen Sirianni and Lewis Friedland in their essay on the subject, "Participatory Democracy in America":

> Civic environmentalism has emerged in recent years as the limits of top-down regulation have become increasingly apparent, and as citizens have continually refined the practices of participatory democracy and collaborative problem solving. In 1970, when the National Environmental Protection Act went into effect, the problems of command-and-control styles of regulation were not well understood, nor did there exist significant institutional capacities for an alternative civic approach. By the 1980s, the problems were quite well understood and, indeed, had become highly contentious in American politics.

Somewhere between the 1970s and the mid 1980s, as Sirianni and Friedland note, it became clear that top-down approaches might work with "point-source pollution" of air and water, but not so well with non-point-source pollution, ecosystem management, pollution prevention or leveraging private incentives for public gain. People got better at tackling such problems on their own, and civic environmentalism was born as a complement to regulation.

Or, more accurately, it was reborn.

Participatory environmentalism is not really new. The writings of James Madison and Alexander Hamilton offer evidence that the founders intended for citizens to exercise their rights and to carry out their duties and responsibilities within the framework of a republic. The founders could not anticipate every problem, but they wrote a recipe for active citizenship that can cook up solutions to today's environmental problems.

Professor Marc Landy of Boston College sees Thomas Jefferson's thinking about the farm as a metaphor for today's movement to restore a sense of balance.

Jefferson's farm provides a middle ground between beautiful but useless nature and corrupt urbanity. His farm is a product of intelligence and ingenuity, not simply a mystical bond with the soil. Yeoman farmers encouraged the moral characteristics that support democracy—patience, resourcefulness and love of order. The tasks of cultivation, shepherding and husbandry proclaim responsibility, thoughtfulness, dutifulness—the important traits of citizenship.

Frederick Law Olmsted, the father of American landscape architecture and the first supervisor of New York City's Central Park, represents another icon for environmental civics—the park. "His great legacy," Landy observes, "consists not only of the parks he designed and the natural wonders he helped to preserve, but also the subtle and complex thinking and writing he produced regarding the relationship of people and nature." Like Jefferson, he understood the importance of place and the influence of physical surroundings on democratic life. In Olmstead's view, "civic friendship" is rooted in a sense of place, and it fosters citizenship across a broad spectrum of public issues and concerns.

Frank Lloyd Wright, the often controversial American architect, also practiced his own brand of environmental civics by insisting that buildings and homes conform with their natural settings. The marriage of human need and natural preservation characterized much of Wright's work, including Taliesin, his home and studio near Spring Green, Wisconsin. Like other architects of the "Prairie School," Wright was suspicious of cities but he wasn't above cutting down a tree or two if it suited his architectural needs. Still, his belief that architecture should be indigenous lives on and challenges today's builders and developers to think about how their work conforms with the land.

Former U.S. senator Gaylord Nelson, now a counselor to the Wilderness Society, helped write the federal environmental legislation passed in the early 1970s. But even Nelson didn't want or expect a bureaucracy-driven environmental movement when he created "Earth Day" in 1970. Instead, he wanted people to get involved—and politicians to respond.

"While the public was concerned about what was happening, the political establishment was not," Nelson wrote in an essay for the *Wisconsin State Journal* (December 31, 1999). "The purpose of Earth Day was to capitalize on this concern by organizing a national, grassroots demonstration so large it would literally shake the political establishment out of its lethargy and force the environment onto the national political agenda.... Finally, the environmental information age evolved."

It is within the writings and actions of Aldo Leopold, however, that a practical foundation for modern civic enviromentalism may be found. Leopold believed that conservation is too important to be left to government alone; it is a realm for individual responsibility, good science and economic reality. Leopold also argued that when there is a harmony of land and owner—when both are improved by reason of their coexistence—there is conservation. The better off people are, the more they will take care of their land. The better the condition of that land, the more it can return value to owners, neighbors and the community at large.

Leopold (1887–1948) lived before the word "environmentalism" was widely used, but his career embodied this term in ways that were ahead of his time. "In my view, [Leopold] was the exemplar of civic environmentalism," said Curt Meine, a Leopold biographer and conservation biologist who has worked with the International Crane Foundation and the Wisconsin Academy of Sciences, Arts and Letters. "He understood that all the efforts in the public sphere would go for naught unless you have a committed, incented and responsible citizenry. He understood that there is a role for government, but it first has to do with where people live and what values they hold. Washington and statehouse politics may be important, but it's all soulless unless you have people doing things on the back forty."

Leopold knew that responsible behavior by one group of people could be contagious for others and that incentives, not regulation, were more likely to produce lasting change. Environmental endeavors, undertaken within a broader civic context, offer opportunities to monitor and represent to others the results of constructive civic behavior. Civic environmentalism is more than a feel-good attitude about the local park. It demonstrates conservation that improves safety, lessens the risk of catastrophic environmental damage, and reduces economic burdens. Best of all, it provides examples that can be passed on from community to community, which are free to shape their own solutions as they see fit.

Is it easy to practice participatory environmentalism? Of course not. It is messier, more time-consuming, more patience-testing and more complicated than an order from on high or a unilateral action by an individual landowner. Forming constructive partnerships is hard and sometimes frustrating work, and not every experiment in participatory environmentalism will succeed. But the effort to do so is essential. As Leopold once said, "We shall never achieve harmony with land, any more than we shall achieve absolute justice or liberty for people. In these higher aspirations the important thing is not to achieve, but to strive."

In time, growing public awareness of the limitations of the old top-down environmentalism and the possibilities of the new participatory environmentalism will generate a natural evolution in environmental thinking. Those limitations include the high costs of a centralized bureaucracy, the inefficient and sometimes painfully slow way in which decisions are made and implemented, and the lack of support from a public that believes it has never truly been consulted. Contrast this with the opportunity for private and community-based action, undertaken at a pace with which owners and other stakeholders are comfortable, and implemented without the costs and red tape that come with the command-and-control model.

Participatory environmentalism emanates from and builds on Aldo Leopold's central tenet: Given incentives to do good for the land, property owners and communities can also do well for themselves. As a result, it is dedicated to creating "win/win" situations in a world that all too often demands that someone lose.

3

Thomas Malthus, Guru of Gloom

The origins of environmental pessimism

Two major environmental movements may share a desire to protect the world's natural resources, but their bedrock values and methods are light years apart. Those seeking to understand today's philosophical and operational differences between these movements must start with Thomas Malthus, an economist who saw every glass as half empty.

ONE OF THE WORLD'S MOST RENOWNED PESSIMISTS was, by all accounts from his own time, a pretty upbeat guy. Thomas Robert Malthus (1766–1834), the economist whose *Essay on the Principle of Population* influences environmental thought to this day, was born into a genteel family south of London, was liberally educated and enjoyed a long and happy marriage. And, despite his conviction that overpopulation would eventually consume all of earth's resources, he fathered three children.

He was an "amiable and benevolent man," wrote one contemporary of Malthus. Another described him as "tall and elegantly formed," adding that "his appearance, no less than his conduct, was that of a perfect gentleman." His admiring students called him "Pops." In temperament and demeanor, Malthus was much less akin to Ebenezer Scrooge in *A Christmas Carol,* Charles Dickens' classic of the same era, than to Scrooge's nephew Fred. Yet the name of Thomas Malthus will forever be associated with an outlook on life that is gloomier than any of the ghosts that tormented Uncle Scrooge.

The enduring pessimism of Malthus is the intellectual cornerstone of the environmental philosophy that dominated the last three decades of the twentieth century. He is the godfather of command-and-control political environmentalism.

Malthus wrote his pamphlet *An Essay on the Principle of Population* (1789) because he couldn't stand listening to the utopian utilitarians of his day, folks who believed that population growth was and always would be an unquestioned asset. Something about their unbridled optimism didn't sit well with his realist mindset, and so, some twenty-five years after Adam Smith wrote his free-market bible, *The Wealth of Nations,* Malthus resolved to show that there are logical limits to how many people the planet can support.

The core of his *Essay* can be found in this sentence (which, by the way, he never supported with scientific fact or anything other than his own assertions): "Population increases in a geometric ratio, while the means of subsistence increases in an arithmetic ratio." In other words, population growth will inevitably outstrip the earth's ability to produce enough food, to supply enough clean air and water, to provide enough land, and to yield enough minerals, timber and other resources. It's only a matter of time, Malthus said, before we grow our way right out of survival. "The power of population is infinitely greater than the power of the Earth to produce subsistence for man," he wrote.

Actually, he conceded, there are two things that keep population down: the necessary evils of vice and misery, whose grim agents are war, famine and disease. When Malthus rewrote and expanded his *Essay* a few years later, he added "moral restraint" as a further check, urging that people delay marriage for as long as possible and not procreate as freely. (So much for those three Malthus kids.)

Time has proven Malthus wrong. Over the past fifty years, agricultural productivity has grown by such astounding proportions that poorer-quality farmland in the developed world can be retired from use for crops. The world food yield per acre has nearly doubled, and food prices have declined as a percentage of family income. Americans spent about half of their incomes on staying fed in the early 1900s; it was about 10 percent in the late 1990s. Scientists now believe the world could feed another one billion people with existing know-how and farm capacity (a number that would be even larger if agricultural biotechnology were allowed to reach its potential). Enough food is available to provide more than four pounds per person or 2,700 calories per day worldwide; but sadly, political despots, corrupt or bumbling bureaucrats and a weak infrastructure in the developing world keep much of it from getting into deserving mouths.

The writings of Malthus were hardly an inspiration to go forth and change the world. If anything, Malthus was telling society: "Give up now; you're wasting your time." For some reason, this personable and literate English gentleman never factored into his equation the intangible human traits of initiative, ingenuity and creativity. Nor did he take into account the power of people coming together to manage their own resources and chart their own futures, as Canadian lawyer and essayist Peter Landry (www.blu-pete.com) has remarked:

> The most grating conclusion of the several which Malthus comes to in his *Essay* is not that eventually population left unchecked will outstrip man's ability to live on this planet (as true a proposition today as it was in 1798); or that war, pestilence, and alike were natural checks against population (they are); but rather that we are all left with a Hobson's choice, with nature being the stable keeper. Or, if one likes, two choices with no difference in the result; either leave the old checks in place (as if we could remove them) or suffer the consequences of overpopulation. It is clear from a reading of his writings that Malthus thought there is nothing we might do to help ourselves; indeed, any law aimed at the betterment of society, to alleviate want and misery, was likely only to aggravate the evils it sought to cure. The only thing for us, is to have faith that the same forces which brought man to his modern state, might be allowed to continue to preserve him.

In his time, Malthus and his theory were as popular as gout. "For 30 years it rained refutations," wrote biographer James Bonar. "Malthus was the most abused man of the age, put down as a man who defended smallpox, slavery and child murder, who denounced soup kitchens, early marriage and parish allowances; who had the impudence to marry after preaching against the evils of a family; who thought the world so badly governed that the best actions do the most harm; who, in short, took all romance out of life."

In reality, Malthus left it to others to wring all the romance out of life. He just gave them the intellectual platform to do so. Today—more than two hundred years after Malthus articulated a theory that failed to foresee the Industrial Revolution, the Green Revolution, technology advances and the simple ability of people to wisely and judiciously manage their own affairs—there are still adherents to his apocalyptic vision. Malthusian thought has proven itself more durable than Malthus himself would ever have imagined, given that he probably assumed humans would have eaten themselves out of house, home and planet by the turn of the millennium.

Two centuries of being wrong has not discouraged the professional doomsayers who are the intellectual heirs of Malthus. Undaunted by the fact that his formula for agricultural Armageddon has failed miserably, political environmentalists have adopted it as one of the philosophical underpinnings of their movement. It's one thing to grow more corn or wheat, they reason, but the world can never produce more natural beauty or resources. Even if incentives and ingenuity have combined to feed more hungry people, these forces cannot replace natural resources once they are lost. Overpopulation, consumerism and resource depletion will be our undoing, the Malthusians conclude with a certainty unfazed by what they see around them.

In their book *Free Market Environmentalism,* Terry L. Anderson and Donald R. Leal explained the assumptions of the Malthusian outlook:

> The feeling that markets and the environment do not mix is buttressed by the perception that resource exploitation and environmental degradation are inextricably linked to economic growth. During the Industrial Revolution in England, [Malthus] articulated this view, hypothesizing that exponential population growth would eventually overwhelm productivity growth and result in famine and pestilence. At the heart of Malthus' logic: population and growing consumption must eventually run into a wall of finite resources.... Modern-day Malthusians have given such dire predictions an aura of credibility by using complex

computer models to predict precisely when Malthusian calamities will occur.

Computer models notwithstanding, they're still wrong. In 1968, Paul Ehrlich wrote *The Population Bomb* to predict that the end foretold by Malthus was nigh. In 1972, the Club of Rome published *Limits to Growth,* which predicted that the world would run out of gold by 1981, mercury by 1985, tin by 1987, zinc by 1990, oil by 1992, and copper, lead and natural gas by 1993. In the late 1970s, a team of scientists at the Massachusetts Institute of Technology predicted that an "uncontrollable decline" in industrial output, food supplies and population would begin in 2005. In 1980, the malaise-stricken administration of President Jimmy Carter presented its "Global 2000" report, which predicted: "If present trends continue...the world in 2000 will be more crowded, more polluted, less stable ecologically, and more vulnerable to disruption than the world we live in now. Serious stresses involving population, resources and environment are clearly visible ahead."

John Kenneth Galbraith captured the prevailing attitude of the 1960s and 1970s in *The Affluent Society,* where he wrote: "The penultimate Western man, stalled in the ultimate traffic jam and slowly succumbing to carbon monoxide, will not be cheered to hear from the last survivor that the gross national product went up by a record amount." This is a modern restatement of the Malthusian mindset: We can have either economic progress or environmental salvation, but never both.

This has proven to be just as wrong as the original premise of Malthus. Free-market economic growth is generally good for the environment, not the other way around. Affluence and improved technology can lead to a cleaner world when citizens use their wealth and know-how to carry out a conservation ethic that preserves a better world for the next generation.

In a February 2002 report for PERC (Political Economy Research Center), the Montana-based center for free-market environmentalism, Professor Seth W. Norton of the Wheaton College school of business took a different approach to assessing what has become known as the "Malthusian population trap." Norton accepted, for the purpose of analysis, that population growth has adverse effects that can be severe. But he examined those effects against the positive influences of "growth-enhancing institutions"—namely, free markets and the rule of law. He concluded that nations without such institutions tend to have problems associated with population growth, while nations with strong economic, legal and political institutions do not.

"The data...suggest there is no population apocalypse and that changes other than reducing population growth will do more for well-being and the environment," Norton wrote in *Population Growth, Economic Freedom and the Rule of Law.* Specifically, he reached four conclusions:

- Market-enhancing economic institutions lower fertility rates.
- The adverse effects of population growth are small.
- Economic institutions can offset those adverse effects.
- Reforming institutions is far more important than controlling population growth.

The problem, of course, is that a good part of the world still lacks institutions that respect contracts and property rights, or that foster free-market innovation. "Numerous nation-states, for various reasons, resist the kind of reform that would ameliorate population problems specifically and human problems in general," Norton wrote.

Professor Norton is not alone in his thinking. Joseph L. Bast, president of the Heartland Institute, also notes the environmental benefits of free markets:

> Countries with capitalist economies invariably have better environmental records than countries with socialist or communist economies. Environmental conditions are improving in every capitalist country in the world, and deteriorating only in non-capitalist countries.... Air and water quality in the United States and other capitalist nations have steadily improved since the 1960s, whereas developing and socialist countries still experience regular public health disasters because of air pollution and tainted water supplies.

Free markets produce better and cleaner technology. Consider the automobile, for instance. Internal combustion vehicles are the bane of the environmental movement, yet they're becoming cleaner and, with the exception of the trendy sports utility vehicle, more fuel-efficient. Those who long for the days before cars should remember that they replaced a far dirtier form of transportation—the horse, which left huge piles of dung in the roads to spread disease or wash into water systems. The automobile is "certainly an improvement from the incredibly filthy streets and waterways of medieval and Renaissance cities," observed economists William Baumol and Wallace Oates. Also, consider the carbon-monoxide-choked air of Galbraith's prediction. Ambient air pollution levels have been decreasing steadily since the 1970s. Between 1976 and 1997, levels of all six major air pollutants fell significantly: sulfur dioxide decreased by 58 percent, nitrogen dioxides by 27 percent; ozone by 30 percent; carbon monoxide by 61 percent (sorry, Mr. Galbraith) and lead by 97 percent. Measured against economic output, air pollution has declined an amazing 3 percent per year since 1940. "Smog" levels that Paul Ehrlich predicted might kill 200,000 New Yorkers and Los Angelinos in 1973 have declined steadily over thirty years. In fact, the Pacific Research Institute in San Francisco issued a 2001 Index of Leading Environmental Indicators showing that aggregate emission of pollutants measured by the Environmental Protection Agency has dropped by 64 percent since 1970.

For decades, however, environmental activists have traded upon fear by reiterating their prophecies of global environmental apocalypse. With "earth in the balance," environmental goals come first and all other concerns are a distant second or third. Because nothing less melodramatic than the "fate of the earth" is at stake, those who question this paradigm must be either ignorant or selfish, or both. In one public opinion poll after another, respondents routinely say they worry a great deal about environmental pollution and resource shortages.

It should give us pause that none of the worst global nightmares have come true. Worldwide "time bombs" have not blown up in our faces. That's not to say there are not serious environmental challenges to be confronted, or that there are no limits to our physical world. But the global catastrophe predicted by Malthus and the other Horsemen of the Apocalypse has not come to pass.

The late Julian Simon was so confident in the ability of the "ultimate resource" (the human mind) to cope with natural resource constraints that he once offered a simple bet to Ehrlich: If your hypothesis about resource scarcity is correct, he said, it should be reflected over time in rising resource prices. In 1980, Simon proposed that Ehrlich pick five resources and that both men hypothetically buy $200 worth of each. Ten years later, they would see if the $1,000 market basket had risen or fallen in value. Ehrlich selected tungsten, copper, chrome, nickel and tin. When the two men checked the prices in 1990, every commodity selected by Ehrlich had fallen in value—even without adjustment for inflation. After inflation, Ehrlich's five resources were valued at less than $500. Ehrlich paid up, grumbling that it would simply take longer for his scarcity theory to prove itself correct. But he refused to renew the bet with Simon for $10,000.

This story illustrates one of the most powerful trends of the twentieth century: the steady and significant progress in reducing almost every form of pollution and in protecting natural resources. Consider the following from *It's Getting Better All the Time*, by economist Simon and the Cato Institute's Stephen Moore:

• The quality of drinking water has demonstrably improved, thanks to technology, improved purification methods and, most importantly, a resolve not to pollute it in the first place. The percentage of water sources that were judged by the Council on Environmental Quality to be poor or severe fell from 30 percent in 1961 (the furthest back in time for which there are reliable data) to 17 percent in 1974, and then to 5 percent in the late 1990s.

• There has been enormous progress in treating industrial and municipal waste before it enters streams, rivers and lakes. Wastewater plants served only 40 million Americans, or about 22 percent of the population, in 1960. Today they serve about 190 million Americans, or 70 percent of the population.

• In 1994, 86 percent of America's rivers and streams were usable for swimming and fishing, as were 91 percent of lakes, up from 36 percent in 1972. Lakes that were pronounced environmentally "dead" in the 1960s and 1970s, such as Erie and Ontario, are now producing record fish catches.

• The biggest oil spill risk in the United States is not a repeat of the 1989 *Exxon Valdez* disaster, but what we blithely pour down our own drains. Steven Hayward of the Pacific Research Institute reported: "American households pour 1.3 billion liters of oil-based products down the drain each year. In comparison, the Exxon Valdez spilled just over 41 million liters of crude oil into Prince William Sound." By volume, the trend line for oil spills has been down since 1973.

• Solid waste in the United States slightly more than doubled between 1960 and 1990, but recycling rose by 96 percent during the same period. About 70 percent of the physical waste now generated in the United States is biodegradable.

• Our society uses energy more efficiently today than ever before. According to calculations by the National Center for Policy Analysis, "the amount of energy needed to produce a dollar of GNP (in real terms) has been steadily declining at a rate of 1 percent per year since 1929. By 1989, the amount of energy needed to produce a dollar of GNP was almost half of what it was 60 years earlier."

• Although the rate is still too high, the loss of U.S. wetlands fell from about 500,000 acres per year in the 1950s, 1960s and 1970s to about 50,000 acres per year between the mid 1980s and mid 1990s. "The Wetland Reserve Program alone has restored as much as 210,000 acres in some years," former Delaware governor Pete du Pont wrote recently for the National Center for Policy Analysis.

Julian Simon's work on natural resources, however, drew skepticism from environmentalists who wanted to believe that things in the world were much worse than they really are. But one such skeptic, the Danish statistics professor Bjorn Lomborg, did what any good scientist should do: He decided to check Simon's facts and figures for himself. In the preface to his 2001 book, *The Skeptical Environmentalist,* Lomborg explained:

> I'm an old left-wing Greenpeace member and for a long time had been concerned about environmental questions. At the same time, I teach statistics, and it should therefore be easy for me to check Simon's sources. Moreover, I always tell my students how statistics is one of science's best ways to check whether our ven-

erable social beliefs stand up to scrutiny or turn out to be myths. Yet, I had never really questioned my own belief in an ever deteriorating environment—and here was Simon, telling me to put my beliefs under the statistical microscope.

In a project that began in the fall of 1997, Lomborg and ten of his best students at the University of Aarhus painstakingly examined statistical evidence concerning a range of environmental problems. The results were surprising: with a few exceptions, Julian Simon was right. Lomborg concluded that there is far more reason for optimism than for pessimism, and he stressed the need for clearheaded prioritization of resources to tackle real, rather than imagined, environmental challenges.

Not surprisingly, *The Skeptical Environmentalist* has stirred both acclaim and criticism internationally. Some environmentalists have responded with the same kind of gut reaction that Lomborg had when he first read Simon's work. Other scientists, however, came to Lomborg's defense. One such scientist was Anthony Trewavas of the University of Edinburgh's Institute of Cell and Molecular Biology.

"Laws do not exist in biology, but generalizations; there are exceptions to every biological principle," wrote Trewavas in the December 2001 edition of *Nature* magazine. "Extrapolating from the past to predict a doom-and-gloom future has been an industry from Malthus onwards. But the ultimate resource is the creativity and skill of the human intellect; formulating the problem often generates solutions."

Responding to a critical review of *The Skeptical Environmentalist* by Stuart Pimm and Jeff Harvey, Trewavas noted the inconsistency of an environmental movement that is dependent on democratic debate refusing to accept the openness of such a debate. "Democracy needs some people to shout loudly about the problems of the world in which we live, but such claims must be treated critically. That is Lomborg's thesis," Trewavas wrote. "Open democratic debate about conservation policy is essential because there are many calls on public resources. The policies that are decided have to be the best return for money, and the public should vote on the outcome."

Confronted with evidence of environmental progress, much of which is due to free minds and markets, one might suppose that the Malthusians and doomsayers would be in full retreat. But they're not. In fact, their line of thinking seems to be stubbornly imbedded in the Western and American psyche.

Recent public-opinion polls indicate that many Americans are unaware of the environmental progress around them. In March 2001, the Gallup Poll asked: "Right now, do you think the quality of the environment in the country as a whole is getting better or getting worse?" Of the 1,060 adults who responded, 36 percent said it's getting better, 57 percent said it's getting worse, 5 percent said it's the same and 2 percent had no opinion. A poll for *Newsweek* magazine conducted in April 2000 by Princeton Survey Research Associates asked: "Since the first Earth Day was held 30 years ago, how much progress—if any—do you think has been made toward solving environmental problems: major progress, minor progress, or no progress, or have environmental problems actually gotten worse?" Of the 752 adults polled, 18 percent said there has been major progress, 52 percent said there has been minor progress, 7 percent said there has been no progress, 16 percent said the problems have gotten worse and 7 percent had no opinion. (See www.PollingReport.com.)

Why do so many Americans doubt environmental progress that, in many cases, they can see with their own eyes? It's because they are repeatedly told otherwise in the news media, where Malthusians have found jobs as newspaper city editors and television news directors. The few national reporters who dare to write about the Environmental Emperor's lack of clothes (ABC television's John Stossel, for example) are attacked by groups hoping to undermine their credibility. It's the time-honored tactic of guerrilla politics: If you cannot refute the argument, tar the credibility of the source.

There's also a simpler explanation that falls into the "man-bites-dog" category of journalistic decision-making: Covering environmental disasters makes for good copy and even better video; covering environmental progress is quite literally like watching grass grow in a brownfield. "Good news, unfortunately, isn't always news," said Michael De Alessi, former director of the Washington-based Center for Private Conservation, a group that reports on environmental success stories through its *Stewardship Chronicles.*

Also, editors more often hear from environmental activists who want to convince them that the world is falling apart. Consider this prediction from *Come the Millennium: Interviews on the Shape of Our Future,* which was published by the American Society of Newspaper Editors:

> "If the world's population of 5.4 billion continues to grow at the current rate for only 500 years, the earth will hold 25 trillion people, each with about 5.4 meters of space to call his or her own," said Gayl Ness, director of the Population Environment Dynamics Project at the University of Michigan. "Because this kind of growth obviously cannot continue, the future will see interlocking developments in three crucial fields—population control, the search for cleaner energy, and efforts to save the environment from the effects of toxification and global warming."

In the face of pronouncements like this, nonpartisan observers such as Denmark's Lomborg have concluded that "We ought not let the environmental organizations, business lobbyists or the media be alone in presenting truths and priorities." Statisticians, like Lomborg himself, need to be heard; so do demographers.

For example, some demographers believe that an emerging problem for much of the industrialized world is not too many people, but too few. If current population and labor force trends hold in the coming decades, the size of the work force in most economically advanced nations will either stagnate or shrink, according to Peter McDonald and Rebecca Kippen of the Australian National University. They wrote about labor supply prospects in sixteen nations for *Population and Development Review* (March 2001). Labor shortages are less likely to plague the United States because birth rates are still relatively high and immigration is encouraged, McDonald and Kippen said. In many nations, however, severe labor shortages may necessitate "policies capable of arresting or reversing the fall of fertility."

Meanwhile, today's Malthusians argue that command-and-control regulations are the only reason for some of the environmental improvements of the past thirty years. If not for the EPA and regulators in the fifty states, they contend, the air would still be dirty and the water would still be polluted. They say that any relaxation in those regulations will invite a return to the "bad old days."

There's little doubt that federal environmental regulations are here to stay when it comes to protecting resources that flow across political borders—air and water being the best examples. In fact, regulatory actions imposed from the top can work when there is a rational cost/benefit analysis.

Consider laws to reduce pollution caused by automobiles. Regulation is necessary because market forces alone wouldn't solve the problem.

"If there were no mandatory requirement to install pollution control equipment on new automobiles, no individual consumer would rationally choose to install the equipment, because his independent action has a negligible effect on the net level of pollution," wrote Frederick G. Jauss IV in the summer 2001 *Administrative Law Review.* "However, if every consumer acts similarly, then the cumulative effect creates a larger pollution problem. Hence, in the absence of some administrative or collective decision to make pollution control mandatory, there is no rational economic motivation to abate the problem." In short, Jauss concluded, "Regulation enters the picture after a determination that certain events not occurring through the free market should occur, and if necessary, through government intervention."

There may always be a need for regulatory action in cases such as those described by Jauss, in which the benefits to society justify the cost of the regulation. However, not all regulatory efforts are subject to cost/benefit analysis, thanks in part to judicial decisions that have made it difficult for administrative agencies to employ commonsense tests.

It would also be a mistake to rest all environmental action on a structure designed for top-down management of emergencies, or to pretend that Americans haven't learned anything about cleaning up their own mess. Regulations can set standards and act as a check against societal or economic excesses, but they cannot motivate people to reason together. They are not a substitute for innovation, incentive-based decisions and true community action.

"The command-and-control approach has accomplished a lot when you look at the Clean Water Act and point-source pollution," said Ed McMahon, director of the American Greenways program of the Conservation Fund. "But there are easily identifiable disasters, too, such as Superfund, which has become the Lawyers' Full Employment Act." Command-and-control environmentalism may have reached its limits, while community-based approaches are "the future of environmentalism. One of the hallmarks of successful communities is that they rely on education, incentives and voluntary action, not just regulation."

Still, Malthus lives on because his ideas are a rallying cry for a movement that thrives on predicting calamities that rarely come true and passing regulations that never go away. Overindulgence in apocalyptic rhetoric comes with a price, however. It produces cynicism, like the boy whose relentless cries of "wolf" caused those around him to miss danger when it actually came howling. It also skews our sense of perspective. What is a major problem and what isn't? How do we accurately assess risks to our health and surroundings? How do we make choices about the costs associated with those risks?

"When every [environmental] problem is treated as a matter of life or death, one loses the ability to discriminate among threats of various degrees of scope, severity and certainty," writes Steven Hayward of the Civic Environmental Working Group. Participatory environmentalism is an attempt to restore that ability to discriminate—to make informed choices—by allowing people to pursue policies that can be measured locally. The average person is unlikely to see whether national efforts to "do something" about greenhouse gases are having an effect, Hayward reasons, but the same citizen can tell whether trees are growing or streams are becoming cleaner close to home.

Hands-on environmentalism does not assume that every problem is equal in scope or global gravity, or that the maxim of "everything is connected to everything else" is a substitute for making local choices. It is about trusting people to develop their own conservation ethic—based on incentives, values and an inborn sense of self-preservation—and following that ethic to compatible solutions.

Thomas Robert Malthus may have been "amiable and benevolent," but he sorely underestimated the strength of the human mind and spirit. As long as we can collectively and individually reason our way through challenges, we will not only survive, but prosper. And the planet will prosper with us.

4

From Malthus to Muir and Pinchot

The birth and division of the modern environmental movement

Sierra Club cofounder John Muir and Gifford Pinchot, the first head of the U.S. Forest Service, were conservationists influenced by Malthusian thought but not bound by it. In time, a feud between these two giants provoked a philosophical split that continues to separate "preservationists" from regulation-minded "conservationists." It would later fall to a man who admired both Muir and Pinchot, Aldo Leopold, to point to a third and more sustainable way.

JOHN MUIR WAS THE SON OF HARD-DRIVING Scottish immigrants who made the mistake of trying to farm in Wisconsin's "District of the Sands," a land rich in beauty but poor in soil. The lessons that young Muir learned on those fragile but dynamic savannas and prairies would guide him—and perhaps haunt him—throughout a life dedicated to a search for Eden.

Gifford Pinchot was born into East Coast wealth, graduated from Yale University and was further schooled at L'Ecole Nationale Forestière in France. There he became imbued with a conservation ethic that combined a sense of man's mastery over nature with a belief that government knows best when it comes to stewardship of land and water.

Following different paths, Muir and Pinchot launched a modern environmental movement that was right for a rapacious time in American history, and which continues to produce benefits for generations of citizens who enjoy what these two helped save. Unfortunately, some of the philosophical seeds that they planted eventually mutated into offshoots of environmentalism that may not resemble what either intended.

There are few names more revered in American or international conservation than John Muir (1838–1914). He was not quite eleven years old when half of his family sailed from Glasgow for a new world, which young Muir later described as "a glorious paradise over the sea." His father, Daniel, eventually decided to settle in Wisconsin, lured by reports of abundant wheat harvests there. The family wound up in south-central Marquette County, in a place dotted by wetlands, kettle lakes and oak-hickory savannah. To Muir, it was paradise. He recalled that first spring on Fountain Lake Farm: "Young hearts, young leaves, flowers, animals, the winds and the streams and the sparkling lake, all wildly, gladly rejoicing together. Oh, that glorious Wisconsin wilderness!"

In time, the intensive farming of winter wheat wore out the sandy soils on the Muir farm, and the family bought more land in 1857. Young Muir was soon to begin training in botany at the University of Wisconsin, which had been founded a decade earlier, and he eagerly applied what he learned in college on the family's land. The new Hickory Hill Farm was better suited to farming, but Muir's first love was still Fountain Lake. In the mid 1860s, the same man who would go on to preserve millions of acres of national parks and wilderness would try to buy back the part of Fountain Lake Farm that contained his favorite garden. He was rebuffed, and the experience appears to have driven him to save other Edens elsewhere.

Aldo Leopold wrote about Muir's frustration in his essay "Good Oak," which is included in *A Sand County Almanac*. Leopold gives a cultural chronology of sorts while sawing through an eighty-year-old oak felled by a storm. When he reaches the point where the tree is severed, the core of the "Good Oak," Leopold says:

> The saw now severs 1865, the pith year of our oak. In that year, John Muir offered to buy from his brother-in-law, who then owned the home farm 30 miles east of my oak, a sanctuary for the wildflowers that had gladdened his youth. His brother-in-law declined to part with the land, but he (Muir) could not suppress the idea: 1865 still stands in Wisconsin history as the birthyear of mercy for things natural, wild and free.

It was also a "birthyear" of sorts for Muir's long career as a conservationist. Muir did many other things in his life: he was an inventor, a carriage maker, a shepherd, a prolific writer and a world traveler driven by wanderlust. First and foremost, however, he was a preserver of wild things. His experience on the Wisconsin farms and, later, his observations on overgrazing in California's Yosemite and Sierra Nevada mountains led him to conclude that man, if given the chance, would ruin just about anything in nature.

Like Malthus, Muir worried that humankind would eventually consume its way into oblivion, as he wrote in *A Thousand-Mile Walk to the Gulf*:

> The world, we are told, was made especially for man a presumption not supported by all the facts. A numerous class of men are painfully astonished whenever they find anything, living or dead, in all God's universe, which they cannot eat or render in some way what they call useful to themselves.... Now, it never seems to occur to these far-seeing teachers that Nature's object in making animals and plants might possibly be first of all the happiness of each one of them, not the creation of all for the happiness of one. Why should man value himself as more than a small part of the one great unit of creation? And what creature of all that the Lord has taken the pains to make is not essential to the completeness of that unit—the cosmos? The universe would be incomplete without man; but it would also be incomplete without the smallest transmicroscopic creature that dwells beyond our conceitful eyes and knowledge.

Muir's conviction led him to save land before the plow or the cow could devastate it. His writings and work led to the creation of national parks at Yosemite, Sequoia, Mount Rainier, the Grand Canyon and the Petrified Forest. He and his associates founded the nation's first environmental group, the Sierra Club, in 1892 to "do something for wildness and make the mountains glad." It was a joyful mission statement that characterized his lifelong sense of awe about nature. Muir is known today as the "father" of the national parks system, and his 1903 conversations with President Teddy Roosevelt, with whom he spent several days alone in the California wilderness, influenced a young chief executive to embark on a program of conservation.

To Muir, the prime directive was to preserve, even if it meant denying man the tangible economic benefits of the land. It was a philosophy that put him on a collision course with a younger contemporary, Gifford Pinchot.

Pinchot (1865–1946) was born the year Muir tried to buy back the Fountain Lake Farm in Wisconsin. Part of his privileged upbringing included stays at a family estate in Pennsylvania called "Grey Towers," a place where Pinchot fell in love with forests. His time at L'Ecole Nationale Forestière (founded by Napoleon to ensure a sustainable supply of timber for the French fleet) persuaded him that a bustling United States needed to do a much better job of managing its forests for the national good. "When I got home at the end of 1890...the nation was obsessed by a fury of development," Pinchot wrote. "The American Colossus was fiercely intent on appropriating and exploiting the riches of the richest of all continents."

Not that Pinchot thought this was entirely a bad thing. Instead, he believed that in order to stave off Malthusian disaster for as long as possible, mankind must do a better, more scientific job of sustaining its resources while simultaneously exploiting them. "The greatest good of the greatest number in the long run" became Pinchot's guiding principle.

Pinchot's attitude was consistent with his time. It was an era when Americans were putting more trust in science, experts and the power of the nation-state. As a political progressive, Pinchot believed that government intervention could temper the excesses of capitalism and that science would lead to a "wise use" of natural resources that would otherwise be depleted.

In 1898, Pinchot was appointed chief of the Division (later the Bureau) of Forestry in the U.S. Department of Agriculture, where he endeavored to protect the forests he loved while ensuring their continued use by a growing nation. In 1905, the bureau was given control of the national forest reserves, and later was renamed the Forest Service. "President Roosevelt, a fellow Republican whom Pinchot greatly admired, allowed him considerable independence in the administration of the service. Pinchot in turn imparted to his staff a spirit of diligence and a sense of mission," reads a Pinchot biography on the Grey Towers National Historic Landmark website (www.pinchot.org).

With Pinchot leading the way, the notion of scientific "wise use" flourished and the national forest system grew from 32 forests in 1898 to 149 with a total of 193 million acres by 1910, when President William Howard Taft finally fired the often-headstrong Pinchot for repeated insubordination. But Pinchot left another legacy that continues to influence environmental policy today: he was the father of command-and-control environmentalism.

A relentless administrator, confident in his ideas, Pinchot built a federal bureaucracy that was both efficient and unafraid of trampling on private property rights if it led to the "greater good." The ends more than justified the means to Pinchot, who believed that the pace of exploitation in his day would bring on disaster sooner rather than later. In this he was similar to Muir, who also believed in the "wise use" theory and in leveraging the power of the state to protect natural wonders.

Where Muir and Pinchot parted ways was over the principle of preservation, as Susanna B. Hecht explained in a February 1993 lecture at the University of California at Berkeley.

> [Pinchot] was concerned to control unbridled excesses of petty entrepreneurialism that characterized the timber industry at the turn of the century. Thus using the emerging, unprecedented regulatory power of the state, he set his task to one of large scale state management of resources, and the creation of a regulatory apparatus focused on forest management along industrial lines. By rationalizing production, wasteful activities would be transcended. Consolidation into state holdings as well as corporate structures would provide economies of scale and economic buffers able to protect (at least in theory) land resources from mining and overuse. For Muir, destruction of resources and their preservation represented a moral crisis, while for Pinchot the problem was merely managerial, a technical one of efficient resource use.

Muir liked Pinchot and spent time with the younger man. But eventually their conflicting outlooks would produce an irreversible split. The immediate issue was the Hetch Hetchy Valley.

In 1906, San Francisco was destroyed not so much by the great earthquake as by the fires that followed in its wake. There wasn't enough water to fight the fires, and the city's leaders then looked to

California's mountains for a steady supply. A plan was developed to dam the Hetch Hetchy Valley, a smaller twin to the Yosemite Valley, and create a reservoir for a growing and fire-scarred city.

The preservationist Muir vehemently opposed the idea. Pinchot, after some initial misgivings, supported it. Although the records aren't altogether clear, it appears that Pinchot eventually fell back on his belief that resources should be used for the "greatest good of the greatest number in the long run." Tourists and naturalists might glory in the majesty of the Hetch Hetchy Valley, he reasoned, but their interests paled in comparison to those of a thirsty metropolis: "The delight of the few men and women who would yearly go into the Hetch Hetchy Valley should not outweight [sic] the conservation policy, to take every part of the land and its resources and put it to use in which it will best serve the most people."

The political battle over Hetch Hetchy ended in 1913, when President Woodrow Wilson signed the bill to construct the dam. Muir died a year later, having not spoken to Pinchot since the latter announced his support for flooding the valley.

The rift between preservation and state-managed conservation grew wider over time. An aggressive wing of the preservation movement drifted away from Muir's notion of preserving special, pristine places and toward the ideal of preserving everything, regardless of cost to the economy or society. Meanwhile, part of the utilitarian conservation movement migrated from the "wise use" concept to a more bureaucratic approach, in which "no use" was the rule unless the right government agencies could be persuaded otherwise. In both cases, private property owners were left out of the equation.

Nearly ninety years after the Hetch Hetchy feud, it is the thinking of Aldo Leopold that holds the power to reunite the best of Muir and Pinchot—and offer a more rational path for today's environmentalists. Leopold admired both men. A forester himself, Leopold was schooled at the Yale forestry school founded by Pinchot in 1900. He devoured Muir's writings and, like Muir, drew heavily on the thinking of Henry David Thoreau, Ralph Waldo Emerson and the Bible.

Early in his career, however, Leopold could best have been

described as a Pinchot disciple. But he came to question Pinchot's principle of "the greatest good of the greatest number in the long run" because he believed it failed to encompass a full understanding of the land community's overall health. Leopold broadened Pinchot's concept of the "greatest good" to include the biotic community itself, as Marybeth Lorbiecki writes in *Aldo Leopold: A Fierce Green Fire.*

> Pinchot, high on the knowledge of the emerging science of forestry, believed that foresters had the information they needed to understand how a forest works. And based on this knowledge and his guiding principle, they could make sustainable land use decisions. However, once Leopold acknowledged that he could not, nor could any human at this point, know the land community well enough to take a beneficial dictatorial position to it, he had to back away from Pinchot's approach—as Pinchot might have done himself had he lived long enough and seen the kinds of things Leopold had seen. Leopold comes in the end to the conclusion that ethics will have to guide us, more than economic considerations or short-sighted self-interest.

On the other hand, Leopold did not entirely agree with Muir. Leopold viewed man as a part of the natural community, not purely an intruder. He saw man as an affectionate partner with nature. "When we see land as a community to which we belong, we may begin to use it with love and respect," Leopold wrote. He also believed that "a land ethic changes the role of *Homo sapiens* from conqueror of the land-community to plain member and citizen of it. It implies respect for his fellow-members, and also respect for the community as such."

As biographer Curt Meine wrote in his 1988 book, *Aldo Leopold: His Life and Work,* Leopold spent much of his life trying to reconcile the conflicting views of Muir and Pinchot:

> The tension between the utilitarian and preservationist views has always existed, not only in society at large, but within the individual as well. Aldo Leopold was a special case in point. Like most foresters, he was drawn to the profession because it allowed him to work with the things he enjoyed most in the place he enjoyed most.... In part, because he held both a Muir-like appreciation of nature and a Pinchot intent to use nature

wisely, Leopold was destined to lead a life of conflicting desires, constant questioning and unending effort to better define the meaning of conservation.

One of the last letters that Leopold wrote before his untimely death in April 1948 was to a Wisconsin conservation leader on the idea of buying one or both of the Muir family farms and preserving them as memorials, a story recounted by Erik R. Brynildson in his 1988 essay for the *Wisconsin Academy Review,* "Restoring the Fountain of John Muir's Youth." Leopold wrote this letter on the day he learned that Oxford Press would publish *A Sand County Almanac.*

A week later, Leopold was gone, but his work lived on. He had finally melded the best of Muir and Pinchot into his own philosophy—an ethic that is not strictly preservationist or utilitarian, but guided by a belief that man can help shape a better world with his own hands.

5

Aldo Leopold and the Origins of "Hands-On" Environmentalism

Long before the word "environmentalism" was coined, conservationist Aldo Leopold advocated a land ethic that relied more on incentives and individual action than on red tape and bureaucracy. His teachings are being rediscovered and embraced by a "hands-on" environmental movement eager to chart a renewed democratic course.

ALDO LEOPOLD BELIEVED IN CONSERVATION and in people's capacity to become interested in and considerate of living things in the world around them. Throughout his long career, he dedicated himself to proving that people—acting individually or collectively, and given proper incentives—could do more good for their natural surroundings than rows of statute books, stacks of legal briefs or roomfuls of bureaucrats. Leopold's emphasis on conservation and participatory environmentalism fell out of favor in the late 1960s and early 1970s, but today his beliefs and the example he set are at the center of a welcome revival in environmental thinking.

The credo of the renewed "hands-on" environmentalism parallels the four central themes of Leopold's work:

- Interaction with the environment has ethical implications.
- Successful conservation efforts ultimately depend on individual responsibility rather than the exercise of governmental largesse and power.
- Conservation efforts are strengthened when they harness the powerful, available synergies of the free market, including

entrepreneurial spirit and a consumerism inspired by doing what's right by nature.

- There is a holistic connectivity within the natural world.

"To preserve the integrity, stability and beauty of the biotic community..."

LEOPOLD'S CONVICTION THAT HUMAN interaction with the environment must have ethical dimensions was rooted in his own experiences and cultivated on a worn-out tract of land in Wisconsin in the early twentieth century.

His land ethic was the logical product of a lifetime of conservation work that melded his love for nature with his appreciation for good science, market incentives and the power of people to leave things better than they found them. "Examine each question in terms of what is ethically and esthetically right, as well as what is economically expedient," Leopold wrote at the end of *A Sand County Almanac,* his best-known work. "A thing is right when it tends to preserve the integrity, stability and beauty of the biotic community. It is wrong when it tends otherwise."

He tested all those principles on land that is now part of the 1,500-acre Leopold Memorial Reserve in Sauk County, Wisconsin. In 1937, Leopold bought an abandoned farm—its soil exhausted from years of intense cultivation, erosion and logging—for a retreat from his academic work as the nation's first professor of wildlife management at the University of Wisconsin in Madison. There he spent weekends with his family in an old cabin he called "the shack," and set out to prove that the land could be nursed back to biological health. Leopold and his family planted hundreds of trees and tested other conservation ideas that bore fruit, figuratively and otherwise.

This experience convinced Leopold that "individual responsibility" is the best starting point for effective land stewardship. As a natural result, he began to question the growing trend in the United States to rely solely on government action for conservation. "We tried to get conservation by buying land, by subsidizing desirable changes in land use, and by passing restrictive laws. The last method largely failed; the other two have produced some small samples of success."

For Leopold, however, private conservation made more sense, and he challenged "the growing assumption that bigger buying (of public land) is a substitute for private conservation practice." Instead, he argued, successful conservation would spring from efforts "rewarding the private landowner who conserves the public interest." Give landowners an economic stake in preserving or creating wildlife habitat, Leopold explained in his essay "Conservation Ethics," and good things usually will happen.

Rejecting the Meadowlark Mentality

LEOPOLD STOOD RELATIVELY ALONE among the major conservation philosophers of his time in arguing that individuals or groups acting in behalf of their own interests, communal or economic, are more effective than government acting alone. In the midst of the Great Depression, a New Deal administration searching desperately for quick fixes to address the very real human needs of the moment gave birth to a school of conservation thought that centered on increasing public stewardship.

Leopold recognized the importance of linking conservation to entrepreneurship and using business tools (contracts, marketing and financing mechanisms) to preserve land and resources; but he was not the Adam Smith of conservation. He was not content to let laissez-faire economic forces chart an uncertain course for nature. In fact, there were many times in his life when he stewed over the mistakes and limitations of capitalism. Yet he appreciated the power of the markets to unleash good as well as evil, and he admonished people to maintain a healthy respect for that power.

"When one considers the prodigious achievements of the profit motive in wrecking land, one hesitates to reject it as a vehicle for restoring land," Leopold wrote in *Round River,* a collection from his journals edited by his son, Luna B. Leopold. Here he also expressed frustration that a generation or more of conservation education had yet to instill in the American landowner a sense of preservation along with economic progress. One or the other seemed to be the norm, Leopold feared, but rarely both. "We can all see profit in conservation practice," he remarked, "but the profit accrues to society rather than the individual. This, of course, explains the trend, at this moment, to wish the whole job on the government."

Because he so mistrusted that trend, Leopold inevitably came home to believing that individual landowners—acting in enlightened self-interest—were a better answer. In *Round River,* he drew upon his observations of his hunting dog to make the point.

> I had a bird dog named Gus. When Gus couldn't find pheasants he worked up an enthusiasm for Sora rails and meadowlarks. This whipped-up zeal for unsatisfactory substitutes masked his failure to find the real thing. It assuaged his inner frustration. We conservationists are like that. We set out a generation ago to convince the American landowner to control fire, to grow forests, to manage wildlife. He did not respond very well. We have virtually no forestry, and mighty little range management, game management, wildflower management, pollution control, or erosion control being practiced voluntarily by private landowners. In many instances the abuse of private land is worse than it was before we started....
>
> To assuage our inner frustration over this failure, we have found us a meadowlark.... The meadowlark was the idea that if the private landowner won't practice conservation, let's build a bureau to do it for him. Like the meadowlark, this substitute has its good points. It smells like success. It is satisfactory on poor lands which bureaus can buy. The trouble is that it contains no device for preventing good private land from becoming poor public land. There is a danger in the assuagement of honest frustration; it helps us forget we have not yet found a pheasant. I'm afraid the meadowlark is not going to remind us. He is flattered by his sudden importance.

To Leopold, the revival of the idea of "legislative compulsion" was another meadowlark. What was needed instead, he argued, was an "organic remedy—something that works from the inside of the economic structure." Leopold found that organic remedy in the practice of consumer-driven conservation. It was a theme reflected in his December 1943 essay in *Audubon Magazine* and in his *Round River* journals, where he wrote:

> We have learned to use our votes and our dollars for conservation. Must we perhaps use our purchasing power also? If

exploitation-lumber and forestry-lumber were each labeled as such, would we prefer to buy the conservation product? If the wheat threshed from burning strawstacks could be labeled as such, would we have the courage to ask for conservation-wheat, and pay for it? If pollution-paper could be distinguished from clean paper, would we pay the extra penny? Over-grazing beef vs. range-management beef? Corn from chernozem, not subsoil? Butter from pasture slopes under 20 percent? Celery from ditchless marshes? Broiled whitefish from five-inch nets? Oranges from unpoisoned groves? A trip to Europe on liners that do not dump their bilgewater? Gasoline from capped wells?

Listening to the Wolf

ALDO LEOPOLD SPENT MUCH OF HIS life worrying about the connections between the natural, civic and economic worlds. That thinking began with his conviction that so many elements of the natural world were linked in ways that people could not fully understand. His essay "Thinking Like a Mountain," which he wrote in two days, captured this belief:

> A deep chesty bawl echoes from rimrock to rimrock, rolls down the mountain, and fades into the far blackness of the night. It is an outburst of wild defiant sorrow, and of contempt for all the adversities of the world. Every living thing (and perhaps many a dead one as well) pays heed to that call. To the deer it is a reminder of the way of all flesh, to the pine a forecast of midnight scuffles and of blood upon the snow, to the coyote a promise of gleanings to come, to the cowman a threat of red ink at the bank, to the hunter a challenge of fang against bullet. Yet behind these obvious and immediate hopes and fears there lies a deeper meaning, known only to the mountain itself. Only the mountain has lived long enough to listen objectively to the howl of a wolf.

In "Thinking Like a Mountain," Leopold exorcised a devil that had tormented him for years—how his killing of a wolf had implications far beyond anything he understood. Only the mountain truly knew how eliminating wolves would lead to an overpopulation of deer and, eventually, the denuding of the mountain itself. Like Saul on the road to Damascus, Leopold saw a guiding light.

We were eating lunch on a high rimrock, at the foot of which a turbulent river elbowed its way. We saw what we thought was a doe fording the torrent, her breast awash in white water. When she climbed the bank toward us and shook out her tail, we realized our error: it was a wolf. A half-dozen others, evidently grown pups, sprang from the willows and all joined in a welcoming melee of wagging tails and playful maulings. What was literally a pile of wolves writhed and tumbled in the center of an open flat at the foot of our rimrock. In those days we had never heard of passing up a chance to kill a wolf. In a second we were pumping lead into the pack, but with more excitement than accuracy; how to aim a steep downhill shot is always confusing. When our rifles were empty, the old wolf was down, and a pup was dragging a leg into impassable side-rocks. We reached the old wolf in time to watch a fierce green fire dying in her eyes. I realized then, and have known ever since, that there was something new to me in those eyes—something known only to her and to the mountain. I was young then, and full of trigger-itch; I thought that because fewer wolves meant more deer, that no wolves would mean hunters' paradise. But after seeing the green fire die, I sensed that neither the wolf nor the mountain agreed with such a view.

...Since then I have lived to see state after state extirpate its wolves. I have watched the face of many a newly wolfless mountain, and seen the south-facing slopes wrinkle with a maze of new deer trails. I have seen every edible bush and seedling browsed, first to anemic desuetude, and then to death. I have seen every edible tree defoliated to the height of a saddlehorn. Such a mountain looks as if someone had given God a new pruning shears, and forbidden Him all other exercise. In the end the starved bones of the hoped-for deer herd, dead of its own too-much, bleach with the bones of the dead sage, or molder under the high-lined junipers."

Leopold's thinking about the connections in nature grew into inquisitiveness about civic and economic connections with nature. In the words of biographer Curt Meine, Leopold became more of an "integrated thinker" than many conservationists of his time. "He put the pieces of the puzzle together" and encouraged others to think about how the relationship between man and nature must forever remain a two-way street.

Starting Again

SADLY, LEOPOLD'S WORK ON THE relationship between ecology, economics and community went into hibernation as less integrated thinkers took command of the environmental movement. It was as if an entire generation had skipped over the wisdom of his time and moved straight into the political environmentalism of the 1970s and beyond. Certainly, times had changed from Leopold's most productive years in the 1920s, '30s and '40s, but not so much that the sense of connections between human and natural communities should have been lost.

Today there is a revival of Leopold's elemental thinking in the civic environmental or community-based environmental movement. Perhaps that's because people recognize that Leopold—like themselves—evolved in his feelings about the environment over time. Just as Leopold's thinking was shaped by his experiences, so are citizens in twenty-first-century America re-examining some of their beliefs about what does and doesn't work.

No longer are people content to sit back and wait for one-size-fits-all rules from a government agency. They are forming alliances between private, nonprofit and public interests. They are relying more on science and less on politics. And they are working together to get things done.

The Sand County Foundation in Madison, Wisconsin, is one of the heirs to the Leopold legacy in its work on community-based conservation networks in North America and Africa. The foundation also manages land around Leopold's Sauk County farm in Wisconsin, hosts executive training seminars, builds coalitions to help remove dams and engages in innovative deer population-control strategies. In all its activities, the Sand County Foundation fosters local partnerships built around cooperation, incentives and the Leopold land ethic.

"More and more, people are stepping up at the local level to protect their own sense of place," said Todd Ambs, former executive director of the River Alliance of Wisconsin. Ten years ago, he noted, there were 12 land trusts in Wisconsin. Today, there are 47. While a decade ago there were a handful of groups organized around rivers and watersheds, today there are 80—a dozen of them organized in the last year. These groups are helping protect wetlands, remove small dams to return rivers to their free-flowing state, conserve wildlife habitat and more, all without the threat of government regulation. It's self-regulation tempered by self-interest and the careful explanation of ecological principles to people who can be trusted to respond to information rather than intimidation.

Leopold cautioned us against chasing the easy "meadowlark" of legislative remedies and urged us instead to pursue the more challenging "pheasant" of a land ethic based on science, economic incentives and communal action. It may be simpler to settle for the meadowlark, but the pheasant is far more satisfying.

6

The Words We Live By

The struggle between these dueling philosophies of environmentalism continues today. If you don't believe it, take a look at the mottos, mission statements and strategies defining, dividing and changing today's environmental movement.

OUR NATION HAS ALWAYS LIVED BY MOTTOS, pledges and creeds, from *E pluribus unum* to the Boy Scout oath to the U.S. Postal Service's promise to deliver mail through rain, sleet and snow. The environmental movement is a part of the same American experience—and you can tell a lot about it by its guiding words.

The first Sierra Club motto reflected the group's roots in the mountainous West. *Altiora peto,* Latin for "I seek high places," was inscribed on the club's 1892 seal. It spoke not only to the search for physically "high places," such as the rugged peaks of the Rocky Mountains, but also to the moral and ethical heights that John Muir and fellow conservationists aspired to climb.

Today the Sierra Club motto is somewhat less lofty, and somewhat more apocalyptic in tone. "One Earth, One Chance," the club proclaims to its 700,000 members. This phrase expresses the environmentalist belief that it's now or never for Planet Earth. Action is not only requested, it's required.

The Sierra Club's mission statement captures some of the mystery and fun of environmental activism, but it still conveys a sense of urgency. The organization aims to:

• Explore, enjoy and protect the wild places of the Earth.

- Practice and promote the responsible use of the Earth's ecosystems and resources.
- Educate and enlist humanity to protect and restore the quality of the natural and human environment.
- Use all lawful means to carry out these objectives.

That final point might be called the Sierra Club's "raise hell and sue the bastards" clause—with some pride by supporters and some chagrin by those who believe the club relies too much on litigation and mass political action and not enough on science, partnerships and true community action.

Let's compare the Sierra Club's mission with those of two other leading environmental organizations, Environmental Defense and the Nature Conservancy. Their words provide an important window into the philosophical divide that has emerged in America's environmental debate. Here's the mission statement of Environmental Defense:

> Environmental Defense is dedicated to protecting the environmental rights of all people, including future generations. Among these rights are clean air, clean water, healthy nourishing food, and a flourishing ecosystem.
>
> Environmental Defense will be guided by scientific evaluation of environmental problems, and the solutions we advocate will be based on science, even when it leads in unfamiliar directions.
>
> Environmental Defense will work to create solutions that win lasting political, economic, and social support because they are bipartisan, efficient, and fair.
>
> Environmental Defense believes that a sustainable environment will require economic and social systems that are equitable and just. We affirm our commitment to the environmental rights of the poor and people of color.
>
> As an American organization, Environmental Defense will always pay special attention to American environmental problems and to America's role in both causing and solving global environmental problems.

Key words pop out: science, people, solutions, bipartisan, efficient, equitable, economic, fair. None of these words appears in the Sierra Club mission statement. To the Sierra Club, people aren't people or citizens; they are "humanity"—just another species competing for resources.

The mission statement of Environmental Defense reflects the thinking of environmental leaders such as Fred Krupp, who, as the organization's executive director since 1986, challenged others in the environmental movement to look beyond toolkits that contained only command-and-control strategies and litigation. In a *Wall Street Journal* piece he called for an approach that takes various human needs into consideration: "Environmentalists should recognize that behind the waste dumps and dams and power plants and pesticides that threaten major environmental harm, there are nearly always legitimate social needs.... The American public does not want conflict between improving our economic well-being and preserving our health and natural resources. The early experience suggests it can have both."

Hardliners have opposed this philosophy as "selling out" or "compromising principles," but it's hard to argue with the results logged by Environmental Defense and other, more holistic environmental organizations. Updated tactics can provide new and more powerful ways to achieve environmental ends. As Krupp said about pollution prevention, "economic incentives can prod people to do the right thing in the first place, rather than requiring a complex regulatory system that brings in costly clean-up as an afterthought."

Not all the tools of the new hands-on environmentalism are economic, of course. Creative use of partnerships and associations is essential; so is truly listening to a community and involving it in the search for solutions. Also, there is a need for framing environmental calls to action within comprehensive operational strategies.

Consider the mission statement and supporting framework of the Nature Conservancy, the world's largest private, international conservation group. It has protected 12.6 million acres in the United States and 80.2 million acres elsewhere in the world, basically one partnership at a time. The mission statement is relatively simple:

> The mission of The Nature Conservancy is to preserve the plants, animals and natural communities that represent the diversity of life on Earth by protecting the lands and waters they need to survive.

It comes with a corollary titled "Our Commitment":

> Through sound science, tangible results and a non-confrontational approach, The Nature Conservancy expands the boundaries of conservation to save the Earth's Last Great Places for future generations.

There is also a "Conservation Vision":

> The Nature Conservancy's vision is to conserve portfolios of functional conservation areas within and across eco-regions. Through this portfolio approach, we will work with partners to conserve a full array of ecological systems and viable native species.

And there is a "Conservation Goal for 2010":

> By 2010, The Nature Conservancy and its partners will take direct action to conserve 600 functional landscapes—500 in the United States and 100 in 35 countries abroad. The Conservancy also will deploy high-leverage strategies to ensure the conservation of at least 2,500 other functional conservation areas—2,000 in the United States and 500 in other countries.

These statements are buttressed by a set of strategies, action plans and measurements that not only give the Nature Conservancy a clear direction, but establish a process for engaging people, making choices and assessing risks and results. In short, the Nature Conservancy's "Conservation by Design" approach is simultaneously inclusive and businesslike—a combination that sustains the organization's work.

Much of the nation's environmental movement operates apart from organizations the size of the Sierra Club, Environmental Defense and the Nature Conservancy. What are the principles that guide hands-on environmental groups that may be much smaller, more ad hoc or more specialized?

At the Sand County Foundation in Madison, Wisconsin, an organization that encourages community-based conservation in North America and Africa, there are three such principles:

- Personal responsibility, good science, economic reality and independent review rather than regulation are the most effective basis for forming productive partnerships.

- Practical advice about using nature's own ecological forces to improve the health of the land is the soundest bedrock on which to build relationships with landowners.
- Aldo Leopold's land ethic is the best guide to responsible, productive, environmental decision-making.

Again, key words and phrases emerge: science, economic reality, responsibility, relationships, partnerships, "independent review rather than regulation."

These themes are consistent with those outlined in a 1995 article on "Civic Environmentalism" by Carmen Sirianni and Lewis Friedland, who described civic environmentalism as "an emerging approach, with a variety of different emphases and methods." Sirianni and Friedland listed the following:

- Collaboration among various communities, interest groups and government agencies, often initiated by a period of adversarial conflict.
- Deliberation among and with various communities, interest groups and levels of government about relative risks and costs, democratic and just ways of allocating these, and common values and interests that might help reframe them.
- Communities that share information and best practices horizontally.
- Federal government role that catalyzes local problem solving within a broader regulatory framework, and policy designs that encourage civic education and responsible action.
- Extensive use of non-regulatory tools.
- A focus on improving the real places and ecosystems in which people live and work, rather than mere statistical risk; on pollution prevention; and on the challenges of non-point sources of pollution.

Key words and phrases: collaboration, communities, deliberation, risks and costs, values and interests, best practices, problem solving, non-regulatory, "real places and ecosystems." It's a pattern that repeats itself whenever and wherever hands-on or participatory environmentalism is practiced.

Sometimes people engage in hands-on environmentalism without a formal organization. They may associate for the express purpose of solving a common problem—or avoiding what Garrett Hardin termed "the tragedy of the commons."

In a 1968 article for *Science* magazine, Professor Hardin presented a chilling picture of how a group of herdsmen inexorably destroyed the shared resource on which they all depended. Essentially, Hardin said, the herdsmen kept adding cattle to a shared range until it was eaten down to the dirt. "Freedom in a commons brings ruin to all," Hardin remarked. Some environmentalists, then and now, have taken this to mean that the government needed to tell those cattle ranchers what to do before they destroyed themselves, and that the only answer to the excesses of the free market is regulation.

Elinor Ostrom, a political economist at Indiana University, believes there are alternatives. Her research shows that people, under many conditions, will work together to manage common resources. Lobster fishers in Maine have done so for years in order to avoid a tragedy in their rich offshore commons. "People not only can [manage resources collectively], but they can outperform a government," said Ostrom.

Scholars of the interwoven human and natural aspects of working landscapes—which in academic language are called social-ecological systems—are rightly concerned about sustainability. They find that much of what constitutes natural resource policy fails the sustainability test. Mandates come from far away and are not connected with meaningful information from the people and the land.

Brian Walker and others writing in a 2004 issue of *Ecology and Society* conclude that because it is essential to know "details of the local and regional context," there will have to be "a different approach to resource governance than currently applied...for a sustainable future." In other words, the ability to carry usable ecological resources into the future will depend in most cases on those people closest to the resources taking on real responsibility, showing capacity for monitoring and management, and developing local governance.

Here are characteristics of successful cases of self-governed, common-pool resource management, as adapted from Ostrom's

1990 work, *Governing the Commons: The Evolution of Institutions for Collective Action.*

1. There is a clear definition of who has the right to use the resource and who does not. The boundaries of the resource are clearly defined.
2. Users must perceive that their required contributions for managing and maintaining the resource are fair in light of the benefits received. Rules governing people's obligations and rules about when and how the resource is used are adapted to the local conditions.
3. Most of the individuals affected by the rules can participate in changing the rules.
4. People who violate the rules are disciplined in accordance with a graduated set of sanctions.
5. Local institutions are available to resolve conflicts quickly.
6. External government authorities do not interfere with resource management schemes developed on a local level.
7. Common-pool resource management systems that are part of larger systems are organized as a series of nested enterprises, each level of which possesses characteristics 1 through 7.

Clear, locally crafted rules; a balance of costs and benefits; a sense of fairness and democracy; rights balanced by obligations; a process for conflict resolution and an avoidance of government interference—these are the characteristics of "common pool" associations such as the lobster fishers of Maine. They are strikingly similar in tone and practice to the rules of emerging hands-on environmental groups.

The environmental movement of the twenty-first century will not be built around mottos or mission statements that inspire anger, fear or the impulse to rely on litigation and bureaucracy. They will be built on words such as collaboration, partnerships, relationships, community, solutions, science and economic reality. Most of all, however, they will be built on the most powerful force in environmentalism: people.

7

"Do as I say" versus "Do as we do" Environmentalism

Institutional command-and-control environmentalism reduces citizens to letter writers and unpaid lobbyists, but there's evidence that people can and will do more if given the chance to get their hands dirty working for a cleaner world.

IF YOU CHECK OUT THE INTERNET SITE of the often-irreverent *Grist Magazine* (www.gristmagazine.com), you'll find a hot button slugged "Do Good." Click there and you'll find a series of well organized, crisply written points under "Take action for the environment." The topics include such things as endangered species, cars and trucks, clean air and water. Here, one might presume, can be found hands-on techniques for serious enviros hoping to save some small corner of the world.

What you find instead is a handbook on how to be an unpaid environmental lobbyist. "Send kids to school in clean buses," proclaims one "cars and trucks" action item in *Grist,* which bills itself as providing "gloom and doom with a sense of humor." How can you get little Johnny and Jill out of those smelly diesel buses? Write a letter to your local school board. "Support tax benefits for smart commuters." More letter writing. "Tell Ford it doesn't get a free pass on climate control." Even more letter writing (with even less chance that it will be read). Here's the one "cars and trucks" item that involves something other than a phone call, e-mail or letter: "Pledge to buy an eco-friendly car."

In the "Climate change and clean energy" category, there's

more run-of-the-mill from *Grist:* "Stop the Bush energy plan." "Tell Bush not to abandon Kyoto." "Ask Exxon/Mobil to contemplate the climate." "Save the Arctic refuge." There are two action points in this category that involve something other than a telephone, fax machine or word processor: "Do a home energy audit" and "Change a habit."

To many in the command-and-control environmental movement, "Do Good" boils down to thinking globally and lobbying locally. Environmentalism is less about getting one's hands dirty close to home than taking long-distance political action to block oil wells in Costa Rica, dams in Belize, whalers in Norway and manatee-killing powerboats in Florida. It's environmentalism via e-mail, in which gains are judged less by land saved or water cleansed than by the weight of congressional mailbags.

At least give *Grist* credit for having a "Do Good" navigational button on its Internet site: not every enviro group aspires to personalize its call for action. However, the "Do Good" list reflects a fundamental problem with the political model for environmental action. With its emphasis on asking government to "do something" about everything, it is undervaluing the power of people to act—individually or collectively, with or without government—in their own back yards.

In the world of enviro-politics, the hands-on approach is discouraged except when it can be directed at problems that lend themselves to point-source solutions. The list is fairly short: recycling and beautification campaigns; home, work and lifestyle conservation; and environmentally correct lobbying.

Recycling

PEOPLE LOVE TO BAG UP THEIR aluminum cans, tie up their newspapers, separate their glass bottles and cut up their cardboard with a razor blade. Why? It feels like they're making a difference. In two short decades, recycling has tapped into values that would not just unravel if all the government subsidies went away overnight, and it has generated a new sort of peer pressure. These days, the green equivalent of "keeping up with the Joneses" is a taller, neater stack of recycling bins.

"Recycling is a hands-on environmental activity that almost everyone identifies with easily," said Perry Mesch, president of Associated Recyclers of Wisconsin and the coordinator of recycling efforts in rural Pepin County. "You don't have to wait for someone else to do something," remarks George Dreckmann, who has managed the successful recycling program for Wisconsin's capital city, Madison, since 1990. "You don't have to wait for a law to be passed."

People who recycle believe they're saving both energy and landfill space by taking ten minutes a week to separate their magazines, plastic bottles and beer bottles from the trash. A recent survey in Wisconsin showed that 75 percent of the state's citizens are strongly committed to recycling and 98 percent report recycling at least some of their household waste. There's also substantial government support for recycling: Wisconsin spends more per capita on recycling ($5.83 a year) than any other state except for Pennsylvania ($6.94), according to a 1999 report by *BioCycle* magazine. It's true that the markets for recycled materials are spotty and in some cases glutted; but try to stop people from recycling now and they'll react as if you had stolen a bag of their aluminum cans.

Mesch sees recycling as entry-level environmental action: It's easy to do and can lead to more personal involvement. "Any little disciplined activity makes the next disciplined activity easier to do," he said. When a desire to do good is coupled with economic incentives, Mesch continued, an environmental ethic becomes ingrained in a community. In Pepin County, a rolling, unglaciated area of western Wisconsin, recycling fits into an indigenous land ethic because farmers there have long sought to protect their fragile soils. According to Mesch, it didn't take much to convince those farmers that recycling is good for their business because it means taking less land out of production for landfills. "It's a part of how we live," he said.

Beautification

LADY BIRD JOHNSON, THE WIFE OF President Lyndon B. Johnson, was appalled by the trash that was piling up on the nation's roadsides in the early 1960s, so she launched a campaign to convince people that littering is unsightly and unhealthy. Her antilittering campaign caught on. The nation's highways became noticeably cleaner when people took it upon themselves to stop tossing hamburger wrappers, soda cans and soiled diapers out their car windows.

Today there are people who organize themselves to pick up other people's litter. Adopt-a-Highway programs exist in virtually every state, and they involve thousands of people who bag millions of tons of trash. Why do they do it? They believe they're making a difference. Every plastic six-pack ring removed from the side of a highway could mean another small animal saved from entanglement and death. Every ditch cleared of trash is a cleaner habitat for birds and other wildlife. The results may be as fleeting as the next careless driver, but there are few environmental actions with more tangible results than those filled bags of trash every few hundred feet along a busy highway.

Adopt-a-Highway programs also appeal to that uniquely American desire for association, which the French visitor Alexis de Tocqueville observed in his tour of the young nation in the 1830s. Americans love to pitch in and do things together, and cleaning up Highway 141 with other members of the Elks Club is a great way to do so—especially when they get credit with a roadside sign.

Jim Swinson, a musician from North Carolina, explained how the desire to "make a difference" inspired him to pick up highway litter (http://pamlicojoe.com):

> I guess I wouldn't be too much of an environmentalist if I didn't "Dare To Influence" in a positive, grassroots kind of way. In 1993 I adopted a highway in my neighborhood. I call it the "Pamlico Joe Expressway" though its given name is Moore's Beach Road. I pick up the litter on this road many times a year as a solo act. I use a crab net to scoop up the trash and I cruise on roller blades. I can do a mile in less than an hour if the litter isn't too bad. My purpose with adopting this highway is twofold: 1) I want to make my neighborhood a little nicer and 2) I want other people to know that one person can make a difference. Many civic and nearby neighborhood groups have noticed

that there is just one of me taking care of a lot of highway and have adopted sections of highway on their own.

Nationally, all but a handful of states have Adopt-a-Highway programs. In "Pamlico Joe's" North Carolina, where Adopt-a-Highway has been in effect since 1988, 6,000 groups and 150,000 volunteers take part, saving taxpayers $3 million each year. In Wisconsin, the ten-year-old program has grown to about 41,000 volunteers in 1,500 groups, who comb the roadsides for litter three times a year, often in weather that tests their resolve. In 2000, they removed 385 tons of litter. "It's an important, hands-on way for people to connect to that thing they both love and hate—the roadway," said Dave Vieth, director of highway operations for the Wisconsin Department of Transportation.

Adopt-a-Highway is successful in part because command-and-control environmentalists didn't create it. The concept came from citizens working with resourceful highway engineers who needed to solve the litter problem after the command-and-control fines for littering failed to produce results. Adopt-a-Highway recognizes the value of incentives, such as the roadside signs and the instant gratification of seeing a clean highway shoulder. There's also room for entrepreneurism in many Adopt-a-Highway programs. In about ten states, private pickup firms sell their services to companies that get roadside credit for the lack of litter. A company makes money, jobs are created and the environment is a bit cleaner. And there's plenty of room for volunteers. Everyone wins.

This approach to cleaning up the transportation corridor through mobilized, organized citizens isn't just good for highways. For the past several years, Chad Pregracke and his organization, Living Lands and Waters, have mobilized locals, concerned businesses, and relevant public agencies to clean up part of their back yards—their own stretches of major rivers. For instance, Chad and his group teamed up with an advocacy group called Missouri River Relief to conduct large-scale cleanups of the Missouri.

Work and Lifestyle Conservation

No bureaucrat tells the average homeowner he must buy a more efficient light bulb or refrigerator, or caulk his windows and roll out some more insulation in the attic. No government agency tells the average business owner she must conduct an energy audit of her office, employ "green-built" construction practices, or install timers and controls to cut down energy use. Yet people are doing these kinds of things every day because homeowners and business managers have at least one powerful incentive to do so: saving money by conserving energy. In the process, they're avoiding a bit of wear and tear on the environment.

At home, at work and (sometimes) in their cars, many Americans practice hands-on environmentalism by exercising common sense and putting their dollars where their principles are. The savings associated with living a conservation-conscious life may not seem like much, especially when your neighbor keeps every light in his house on all night, but the statistics are surprising.

The average car releases about one pound of carbon dioxide for every 20 miles of travel; at mileage rates of about 12,000 miles per year, reducing travel or increasing gas mileage by 10 percent can save an average of 60 pounds of CO_2 per year. The average home produces 4.5 pounds of carbon dioxide through electricity consumption daily, so cutting use by 10 percent can save 165 pounds per year. Start multiplying that by the millions of cars and homes in America, and pretty soon you're talking about a Kyoto Protocol minus the obnoxious diplomats.

Although it may take higher energy prices to get the attention of most Americans, many are already doing what they can to save energy, help the environment and keep more money in their own pockets. Here are a few examples of easy-to-accomplish savings:

- A car tune-up saves between 6 and 9 percent on gas mileage; keeping tires properly inflated saves more.
- Think you'll be back in that room in a couple of minutes? You probably won't, so turn out the light. While you're at it, turn off that computer monitor—it uses as much juice as a 60-watt light bulb.
- Buying appliances with an Energy Star label does more than make the manufacturer feel good. Refrigerators, dishwashers, office equipment, windows and more that merit industry and government Energy Star status help pay for themselves by saving energy. A fully equipped "Energy Star" house can trim 30 percent from its power bill.

- Speaking of refrigerators, keeping them at the right temperature (0 to 5 degrees for the freezer and 38 to 42 degrees for the fridge) can cut bills sharply. By the way, water heaters should be set at no more than 120 degrees to save energy. It's safer, too.
- Lighting accounts for 25 percent of America's electricity use. Replacing standard incandescent bulbs with compact fluorescent bulbs may cost more up front, but they last ten times longer on average and save energy. By the way, halogen lamps are grossly inefficient.
- If everyone in the United States bought the most energy-efficient car in the class they would ordinarily buy (for example, the most fuel-efficient sports utility vehicle instead of the least efficient SUV), it would save 1.47 billion gallons of gasoline per year, according to *Grist Magazine*.
- Energy-efficient showerheads can pay for themselves in two months by saving 27 cents per day on water and 51 cents on electricity needed to heat the water.

Citizens and businesses are demanding more efficient appliances, tools, lights and vehicles every day—and industry is trying to supply them, with the help of new technology that will drive stakes into the hearts of the energy vampires that quietly drain us. By practicing consumer-driven conservation, as envisioned by Aldo Leopold two generations ago, Americans are forcing long-term changes that will help the environment and their own wallets.

How many Americans does it take to change a standard light bulb to a compact fluorescent? Preferably, all of them.

Lobbying

THE MODERN "ENVIRO-CITIZEN" IS SEEN as a foot soldier in a political war that has many fronts: the courthouse, the statehouse, the halls of Congress and capitols around the world. Few movements are as adept as the environmental movement at organizing citizens to write, call, e-mail or otherwise lobby government. The network can produce thousands of contacts that collectively bring enormous pressure to bear on elected officials and bureaucrats who, absent better ideas, are inclined to overreact. That can bring more top-down regulation, less private action and, ironically, a style of environmental action that is more detached from the ordinary person.

Political action may work in some cases, but it's much less satisfying to fire off a letter to the Costa Rican ambassador or the president of Exxon than to accomplish something tangible in your own back yard. The civic or hands-on approach to environmentalism tells people they can do more than recycle, pick up litter, set out a bird feeder, vote, give money to the Sierra Club, attend a once-a-year Earth Day rally and write letters. It tells them they can work routinely with fellow citizens, corporations, nonprofit groups and local governments to get things done.

"That's where the energy is now in the environmental movement," said Curt Meine, a conservation biologist with the International Crane Foundation in Baraboo, Wisconsin. Meine has also been active in a groundbreaking effort to return about 7,000 acres of a surplus Army ammunition plant to conservation and recreation use. "It's breaking down the traditional boundaries. It's telling them there are new ways to accomplish things. And it has given people of different backgrounds and political leanings an opportunity to come together around the center."

Unfortunately, hands-on environmentalism can be difficult when the prevailing attitude in the command-and-control model is to tell people to keep their hands off.

The Environmental Protection Agency can be an impediment to participatory environmentalism. The EPA is, by definition, a regulatory agency that exists to enforce laws and administrative rules passed in the hope of creating cleaner air, water and land. That sometimes yields inflexibility in how the rules are interpreted, as well as the rejection of local concerns and alternatives based on sound science. Because local needs are so often ignored, public complaints about the performance of the EPA are on the rise and some members of Congress want to create an ombudsman office within the agency to protect the very public that the EPA is chartered to serve. The absolutism of the command-and-control approach, symbolized by the EPA, shows how political environmentalism veered off track.

"Quality of life resides in pursuit of multiple values," wrote Lynn Scarlett, who was vice president of research for the Reason

Foundation before joining the U.S. Department of the Interior. "People seek shelter, nourishment, health, security, learning, fairness, companionship, freedom and personal comfort together with environmental protection. They even seek many, sometimes competing, environmental goals."

This begs for setting priorities and making choices, for balancing competing community values within the bounds of what's considered acceptable environmental risk. Unfortunately, the regulatory model can stand in the way of that community-based process.

Consider what happened in 1998 when Select Steel Inc. proposed construction of a $175 million steel mill that would create two hundred jobs for Genesee County, Michigan, which includes Flint, a city that was desperate for jobs. Most people in the economically depressed community welcomed it, but the EPA's "environmental justice" policy got in the way. This policy is intended to protect minorities from being disproportionately affected by pollution. Activists in Michigan charged that pollution from the steel plant would unfairly rain down on Flint's black and Hispanic populations. After initiating an investigation, the EPA announced it would drop the case after discovering that the area surrounding the plant was, in fact, 84 percent white. End of story? Hardly. Activists appealed, the investigation dragged on, and nine months later Select Steel decided to build its plant instead in Ingham County, which includes Lansing, the state capital, and where the unemployment rate was 3.2 percent versus 5.6 percent in Genesee County.

The "environmental justice" policy stemmed from a 1994 executive order signed by President Clinton, who gave the EPA power to deny environmental permits whenever it determines that pollutants will disproportionately affect communities with particular racial or ethnic characteristics. What those communities actually want, however, is irrelevant.

Some African-American leaders—those who want to balance economic development with environmental protection—have condemned the EPA's approach. Not only does the policy run counter to federal and state goals of developing "brownfield" sites, it also causes time-consuming delays that deprive inner cities of necessary economic opportunity. "The EPA is pimping the black community to further their own agenda of a pristine earth at the expense of our jobs," said Harry Alford, president of the Black Chamber of Commerce. Detroit's mayor Dennis Archer said the EPA's policy is "so vague and so broad that it nullifies everything we have done to attract companies to our brownfield sites."

Project 21, an African-American leadership network affiliated with the National Center for Public Policy Research, concluded that environmental regulations "unjustly burden minority and low-income Americans." In an April 2002 statement, the group noted EPA estimates that the average American household pays $3,000 per year to comply with environmental laws. "Since black families earn less than white families on average, black families spend approximately 12 percent of their incomes on costs related to environmental regulations while white families pay only 7 percent," Project 21 reported.

Meanwhile, the environmental justice policy can act to keep those same black families from earning more. Writing for the National Center for Public Policy Research (www.nationalcenter.org), Michael Centrone called the EPA's environmental justice policy "an injustice…. Contrary to what environmentalists may believe, African-Americans do not need special protection from economic improvement."

The EPA is not alone when it comes to being tone deaf to the symphony of local concerns. State environmental agencies also get mired down in their own bureaucracy, as a number of Wisconsin cities and villages discovered when they tried to deal with "isolated" wetlands. These are different from "connected" wetlands in that they're not linked to other bodies of surface water, such as streams, rivers or lakes. They're free-standing marshes or, in some cases, low places that collect water in rainy seasons. During a legislative debate on regulation of isolated wetlands in Wisconsin, the Wisconsin Alliance of Cities collected examples of bureaucratic overreaching. Here are two:

• In the south-central city of Beaver Dam, city officials wanted to extend a street to provide a safe bicycling route to a city park. The alternative was for kids to bike along a busy state highway. However, the state Department of Natural Resources (DNR) said no because the street extension would have filled in about two-tenths of an acre of isolated wetland. The DNR said it had jurisdiction because an old drainage ditch from a nearby farm was a "navigable stream."

• In the state capital of Madison, city officials wanted to clean up a 100-foot-wide grass drainage way, just as it had done many times in the past without DNR meddling. This time, the DNR barged in, claiming that creation of a retention pond would interfere with "fish migration." Mused Rich Eggleston of the Wisconsin Alliance of Cities: "The agency did not share with city officials the identity of the fish species that migrates in grasslands."

On balance, however, regulation based at the state level is preferable to federal regulation—if the issues are self-contained within state borders or can be solved through regional cooperation. Simply put, state governments are much closer to the people and, thus, more responsive and willing to experiment.

In *Let Fifty Flowers Bloom: Transforming the States into Laboratories of Environmental Policy*, Professor Jonathan H. Adler of Case Western Reserve Law School concluded that "states are on the front lines of developing new and innovative approaches" to environmental protection.

> There is a general consensus that the current federal environmental regulatory system is broken and needs repair. Current environmental programs exhibit most of the failings of Soviet-style, command-and-control systems: excessive rigidity, inefficiency, diminishing marginal returns, poor prioritization.... The local and regional nature of many environmental problems means that local knowledge and expertise is necessary to develop proper solutions. Such localized knowledge is simply beyond the grasp of even the most intrepid federal regulators.... Returning environmental authority to the states would foster innovation and greater attention to local environmental concerns and conditions, while enhancing accountability for environmental decisions, particularly where environmental concerns are local in nature.

When regulators don't listen to local concerns or when they value process over product, it becomes more difficult to build a sustainable, civic environmentalism that invites the people most affected by those rules and regulations to take responsibility. When regulators work with citizens, business owners and civic leaders in authentic ways (above and beyond nodding their heads in public hearings and then ignoring the testimony), constructive relationships can be forged.

This was President George W. Bush's message in his May 30, 2001 press conference in Three Rivers, California, where he stood beneath a 2,100-year-old giant sequoia and asked that federal conservation efforts show more deference to states, localities and private groups. In his first major speech on the environment, Bush called for a "new environmentalism" that will "protect the claims of nature while protecting the legal rights of property owners." He said, "My administration will adopt a new spirit of respect and cooperation because, in the end, that is the better way to protect the environment we will all share. Citizens and private groups play a crucial role. Just as we share an ethic of stewardship, we must share in the work of stewardship." It remains to be seen whether such a commitment will be manifest in new federal policies that give real encouragement to citizens in practicing an ethic of stewardship.

Hands-on environmentalism means allowing people to take a stake in their own futures, to reason together and to shape solutions that reflect economic reality, scientific fact and ethical choices. People can do more than recycle and pick up litter. All the environmental movement and its regulators need to do is give them the chance.

8

How to Get Your Hands Dirty...and Your Community Clean

A citizens' guide to environmental participation

Many Americans yearn to re-engage in public life, but it may be harder to do so in the institutionalized environmental movement than in other sectors. Here are some rules of (green) thumb for how to get involved in your own hometown.

ACROSS A CONTINENT AND OVER MORE than three centuries, this nation's history has been a study in resiliency and innovation. Refusing to despair in the face of adversity or surrender to doomsday predictions, the men and women of America have continuously found intriguing opportunities and broad new visions in the challenges that come their way.

We're at it again.

In inner-city neighborhoods from coast to coast, cops who in the not-so-distant past rarely ventured out of their squad cars except to buy coffee and doughnuts are now pounding the pavement, talking with citizens who once feared anyone in a blue uniform, and reporting to superiors in decentralized precincts that are located just around the corner.

In public school systems where students are lagging and where learning is compromised, parents aren't waiting for an expert from the state school agency to show up and wave a bureaucrat's magic wand. They're pursuing local options—such as charter schools, private vouchers and more flexible classroom and school management—to help their children get a sound education.

In state capitals from Albany to Sacramento, governors and

legislators who only twenty years ago were content to sit back and wait for the latest pronouncement from Washington are working around federal mandates and tackling problems as complex as welfare reform, health-care financing and energy reliability, often with the help of citizen groups, businesses and other partners outside government. A big motivator for all responsible politicians is to ensure that the taxpayers can afford what is proposed. Budget realities have necessitated reductions in force across a number of agencies in a number of states. This affects the state's capacity to invest in environmental efforts.

In newsrooms large and small, reporters and editors who were trained to find news by thumbing through their Rolodexes in search of elected officials, lobbyists, bureaucrats, special-interest activists and other "experts" are finding new ways to talk with their readers, viewers and listeners. These journalists are discovering that they are writing more thorough, more interesting stories and contributing to the search for solutions rather than reciting problems in the tiresome language of conflict.

Whether these examples are called "community policing," "school choice," "the new federalism" or "civic journalism," they represent a trend that is growing as Americans, confident in their own common sense and blessed with unprecedented access to information, find they can again take control of their lives and their communities.

Scholars who have studied these civic movements in the United States, such as James L. Fishkin, the University of Texas professor who organized the first National Issues Convention in 1996, point to four procedural characteristics that exemplify democratic or civic movements.

Participation—True civic movements seek to maximize individual participation in discussions about the issues and engagement in the activities that affect them, their families and their communities.

Deliberation—Real civic activity is distinguished by efforts to develop the informational capacity essential to making choices.

Political equality—Genuine civic movements provide everyone in a community with a stake and a vote.

Non-tyranny—Civic movements strive for consensus and seek

solutions that serve the best interests of the community and the individual without harming the interests or well-being of either.

The old command-and-control agenda of the environmental politicians, or "enviro-pols," and the new environmentalism grounded in Aldo Leopold's ethos of personal responsibility for environmental stewardship can lay claim to three of Dr. Fishkin's civic movement criteria. They can both claim to advocate participation and political equality. They both support a deliberative process, although the enviro-pols sometimes tend to lose interest in the deliberations when the facts do not support their agenda.

The old and new movements differ sharply when it comes to the fourth criterion, non-tyranny. Explaining why he views non-tyranny as one of the most important criteria, Fishkin wrote: "We could satisfy all our other conditions, the people could speak through their own participation, votes could be counted as to satisfy political equality, the issues could be fully debated so as to satisfy deliberation. Still, the system could result in consequences that destroy the rights or the essential interests or the liberties of some portion of the population, even when the imposition of these deprivations was entirely avoidable."

This should sound familiar to those who have been frustrated by political environmentalism as it has been practiced in the United States for thirty years. Modern enviro-pols follow democratic principles to the letter when they insist on endless process, regulation and litigation, and they will defend the results as democratically achieved. And yet, political environmentalism can be tantamount to a legitimized tyranny when property rights are destroyed and liberties are compromised for some disfavored portion of the population—usually without any need for such "deprivations."

Participatory environmentalism is an antidote to antidemocratic tyranny because it seeks to avoid a majority imposing its will on a minority, especially when there are ways—often through civic deliberation and, in many cases, free markets—to avoid doing so.

Central to the newest wave of American ingenuity is a reassertion of the importance of local decision making. These locally driven decisions are changing institutions, returning power to citizens and providing effective, affordable alternatives to command-and-control strategies that were developed to deal with specific challenges in the last half of the twentieth century, but are now clearly limited and limiting in their contemporary applicability.

In the chapters ahead, we'll take a closer look at community-based, participatory environmentalism at work. You'll take a virtual tour of an America—and a world—where hands-on environmentalists are confronting very different challenges and resolving them in very different places. But these stories are bound together by common threads. Whether it's done in Louisiana, South Carolina, Arizona or Zambia, hands-on environmentalism meets James Fishkin's criteria for participatory democracy; it embodies the values of Aldo Leopold's land ethic; and it aspires to Lynn Scarlett's desire for "a set of institutional arrangements and opportunities that tap local knowledge, foster tailored creativity and innovation, inspire folks to pursue environmental values, create a context for cooperation, and provide decision settings that foster a holistic look at problems, values and opportunities."

The stories that follow illustrate how these principles are being put to work by people who are using their skills and experience along with a conservation ethic to find solutions that work for them, their neighbors and the land. They are the pioneers of hands-on environmentalism.

9

How to Wrestle with a Bear— and Win

The story of Louisiana's Black Bear Conservation Commitee

An Endangered Species Act listing can be something to fear if you're a private landowner, but a group of private conservationists in the Mississippi River Delta found a way to make regulations work for everyone, including the rare Louisiana black bear.

NOT EVERY HANDS-ON ENVIRONMENTAL group sets ground rules this down-home or concise, but here are the "Southern Rules of Engagement" from the Black Bear Conservation Committee in Baton Rouge, Louisiana.

1. Come to the table. The world is run by people who show up.
2. Leave your organizational 2x4 at the door. Polarized opinion generates more heat than light—and has no place at the resource management table.
3. Pick solutions, not fights. Search for the most expansive common ground that is least intrusive.
4. Attack ideas, not individuals. Differences of opinion can lead to enlightened decision-making. No personal attacks. One strike and you're out.
5. Have fun. "We shall never achieve harmony with land, any more than we shall achieve absolute justice or liberty for people. In these higher aspirations the important thing is not to achieve, but to strive."—Aldo Leopold, *A Sand County Almanac.*

These commonsense rules have become a way of life for members of the Black Bear Conservation Committee, which was founded in 1990 to come up with a community-based strategy to protect the rare Louisiana black bear. It wasn't easy and it didn't happen without a lot of hard work, but the success of the Black Bear Conservation Committee (www.bbcc.org) serves as an example of what can happen when people are free to practice participatory environmentalism.

In the late 1980s, the U.S. Fish and Wildlife Service served notice that it would likely put the Louisiana black bear on its endangered species list. The agency had good arguments for doing so. Years of hunting and habitat destruction had taken a toll on the bear, a shy and genetically distinct creature that requires huge areas in which to roam, hunt and live. As best as anyone could tell at the time, there were only a few hundred Louisiana black bears left alive in the Mississippi River lowlands of Louisiana, Mississippi and Arkansas.

Private conservationists in Louisiana and across the Mississippi lowlands wanted to save the bear, but they also realized that a listing under the Endangered Species Act could mean an end to property rights as they knew them. Louisiana farmers, timber companies, environmentalists and regulators—eager to avoid a repeat of the spotted owl fiasco of the Pacific Northwest, where loggers engaged in guerrilla warfare with bureaucrats and environmentalists—resolved to talk about solutions. Because 90 percent of the bear's forested habitat rested in private hands, a private-public partnership was not only possible; it was essential.

What emerged was the Black Bear Conservation Committee. It was an eclectic and even somewhat unlikely collection of citizens, but they embraced a cooperative management approach that turned the typical "lose-lose" story of an Endangered Species Act designation into a "win-win" story for landowners and, most important, for the bear.

"It was tenuous, at best, when we got started. But we were able to pull a lot of people to the table who wouldn't be there otherwise," said Paul Davidson, the BBCC's executive director and a former Louisiana Sierra Club president. "That's because they were all committed to saving bears, for starters, but it was also clear to them that the only way to do so was to come up with a plan that would be amenable to landowners."

Paper companies with vast holdings in the lowlands hardwood forests of the region were threatening to sue the U.S. Fish and Wildlife Service if they put the bear on the endangered species list—and the national Sierra Club was pledging legal action if they didn't. It was within this contentious context that the Black Bear Conservation Committee began to build consensus, piece by piece, relying on solid science, good intentions and some creative use of government incentives and existing law.

A biologist with the Fish and Wildlife Service in Jackson, Mississippi, helped break the logjam by doing something regulators don't often do: He pointed out a constructive loophole in the law. Section 4(d) of the Endangered Species Act could allow timber companies to continue normal selective harvesting of their lands so long as they left undisturbed larger trees with cavities that could serve as bear dens. The timber companies readily agreed, and the provision was written into the Fish and Wildlife Service's eventual listing of the bear.

"That kept the timber companies at the table," said Davidson, whose Louisiana drawl picks up an excited pace when he talks about the progress that cooperation has enabled. "It created an atmosphere that was positive, and it has endured." Davidson credits "very progressive biologists" at Anderson Tully Company, Temple Inland Corporation and International Paper for making it clear that they were hoping to do what was best for the bear. They contributed money for scientific studies of the Louisiana black bear and accelerated their own conservation work, which was already so successful that their lands hosted more migratory and subtropical birds than did reserves on public lands.

The next breakthrough came from an unlikely source: the 1990 Farm Bill. Rewritten every five years or so, federal farm bills are a complicated maze of subsidies and marketing orders. This measure, however, offered a Wetlands Reserve Program that was about to convert the Louisiana black bear from a nuisance to an asset. This program provided cash incentives for the owners of marginal or unproductive farmland with appropriate hydrological features to put their land back into trees. Initially designed to take a million acres of agricultural land out of production in an effort to stabilize commodity prices, the Wetlands Reserve Program had the somewhat unintended effect of becoming a conservation program.

"At the time I don't think many people realized the potential the Wetlands Reserve Program had for restoring wetland habitat in the lower Mississippi Valley. I know that those of us in the bear recovery business had no clue," Davidson recalled.

They found a clue in a hurry when the Black Bear Conservation Committee learned that the new law could pay farmers to reforest marginal lands. "There were literally millions of acres of bottom lands that were too wet to farm" but that farmers felt obliged to plow and plant if they couldn't put the land to some other productive use. That productive use became creating habitat for bears and other threatened species.

Armed with highlighter pens and yard after yard of highly detailed maps, members of the Black Bear Conservation Committee set out to help those farmers apply to the Wetlands Reserve Program. The committee figured out which lands were in black bear "priority areas" and which were not, and mapped out corridors to link unconnected habitats. A point system was developed to compare the relative conservation value of private lands so that limited incentives could be put to best use.

The result: About 350,000 acres of Mississippi River lowlands in Louisiana, Mississippi and Arkansas have been planted in a mix of hardwood trees since 1990, creating new roaming grounds for the Louisiana black bear. Except for selective cutting, these lands must be kept forested in perpetuity.

"The bear became an asset to the average landowner," Davidson said. He now believes there should be an Endangered Species Reserve Program. "What incentives could be provided to landowners willing to manage their private property for the protection, preservation and recovery of the growing list of species that are in trouble? Currently listed or potentially listed species are considered a liability to the private landowner," he said. "How can we turn this around?"

He envisions a program that could include cash payments to some landowners and tax incentives or mitigation banking for others. "I don't pretend to know all the potential incentives that could be incorporated into this program, but I know that the more options provided to potential cooperators, the more cooperation we will have," Davidson said. "Our society provides incentives for almost everything else we want. We give tax breaks to corporate interests in exchange for jobs, get cash rebates for purchasing all sorts of items, and go to work every day because of the financial rewards. If you think about it, almost everything we do is incentive-related. Even our recreational pastimes are incentive-based, the incentive being the pleasure we receive by the experience."

No one knows for sure how many Louisiana black bears there are today, because they shun human contact and rarely call attention to themselves by attacking livestock or otherwise being nuisances. But Davidson estimates that there are between 400 and 500 in Louisiana alone, based on sightings and more unfortunate anecdotal evidence, such as bears getting hit on the highways. "I think we've probably had more bears today in Louisiana than we've had in a hundred years," he said.

Ask Davidson to describe the core reasons for the success of the Black Bear Conservation Committee and he won't dwell on money or regulatory flexibility for long. He will talk a lot about participatory democracy, however. "We, as citizens, have sort of backed off and allowed government to take control. Government in a democracy is not designed to rule like that. It's designed to be a partner," he said. "Communities are supposed to take care of themselves, with the assistance of government."

As the Southern Rules of Engagement so clearly state, the Black Bear Conservation Committee aspires to seat everyone at the table. "That's probably the main reason we don't have anybody throwing rocks at us, because they're all at the table," Davidson remarked. "Our role is to create a forum that keeps everybody working together. None of this is about 'good people' and 'bad people.' It's all about people working together."

The Black Bear Conservation Committee also throws out one of the standard rules of command-and-control environmentalism: the notion that nothing is gained unless everyone gives up something. "If you aim low at a target, that's where you'll hit—low," said Davidson. "We started by listing what we needed rather than by what we needed to give up. We discovered that, by and large, there was nothing we couldn't accomplish."

Davidson had three other pieces of advice for community-based conservation groups:

- *Base things on science.* "I love people who are passionate about things, and I would much rather deal with someone who has a different opinion from me than someone who has no opinion at all," Davidson said. "But you've got to get the facts."
- *Have fun.* "At our meetings, we always have beer. We always have something to eat. We might be able to fish a little bit. We have fun," he said. "Wildlife management is people management. We've got to figure out a way to get people to the table. You can have a wallet full of money, but unless you've got people who want to make sure the water, air and land are clean and are motivated to do so, you don't have anything."
- *Be patient.* Efforts such as the Black Bear Conservation Committee take longer to get organized because there must be genuine participation by all stakeholders. But community-based conservation can save time in the long haul because the product is more likely to stand up under fire. "All the problems we have in resource management could be solved with a more cooperative approach," Davidson said.

As it enters its second decade, the Black Bear Conservation Committee also runs educational workshops for landowners and local communities, provides educational materials to landowners, hunters and others, works with communities to develop management plans, provides educational materials to teachers and generally serves as a model for a cooperative approach to resource management. A notable outreach effort is this committee's use of dogs, trained to keep black bears away from situations that are bad for the bear and for people, as conservation ambassadors too. This approach is being picked up by groups in other states where different subspecies of black bears are coming into conflict with people's abodes, bird feeders, fruit crops, honey hives and the like.

The Black Bear Conservation Committee has made citizens, government, economic incentives and science work together to help protect the Louisiana black bear throughout its low-country home in the lower Mississippi River valley. It's a model that other hands-on environmentalists would do well to follow.

10

"Where There Is No Vision..."

The rebirth of South Carolina'a Cypress Bay Plantation

Just as Aldo Leopold took a worn-out southern Wisconsin farm and turned it into productive land, today's Americans are employing private and community-based efforts to restore landscapes left for dead. Skeet and Gail Burris of South Carolina's Cypress Bay Plantation are evidence that Leopold's legacy lives on.

ALDO LEOPOLD WASN'T ALWAYS SANGUINE about the ability of man to strike a truce with nature. "Why is it that conservation is so rarely practiced by those who must extract a living from the land?" he asked in an essay later published as part of his *Round River* collection. "It is said to boil down, in the last analysis, to economic obstacles. Take forestry as an example: the lumberman says he will crop his timber when stumpage values rise high enough, and when wood substitutes quit underselling him. He said this decades ago. In the interim, stumpage values have gone down, not up; substitutes have increased, not decreased. Forest devastation goes on as before...."

If Leopold were alive today, he would be pleasantly surprised by the recovery of the nation's woodlands. Instead of devastation, there is careful reforestation, often carried out by land trusts, conservancies and private tree farmers who eventually learned the lessons that Leopold wanted them to learn but doubted they ever would. The American Tree Farm System, founded some fifteen years after Leopold's lament, has about 66,000 private landowner members who manage 83 million acres nationwide. Among the best of these landowners are Skeet and Gail Burris of the 1,083-acre Cypress Bay Plantation in South Carolina. Their story is proof that Leopold left an enduring legacy.

The Burrises didn't have to get into the forestry business. Skeet was a successful orthodontist and civic leader in picturesque Beaufort; Gail had her own career and a degree in economics; together they enjoyed their life in the low country of South Carolina that stretches between Charleston and Savannah, Georgia. But there was something about their east Tennessee roots that made them long to become forest landowners. Buying an existing plantation was a daunting prospect due to cost and availability, though starting a tree farm from what amounted to scratch was a possibility.

Skeet Burris finally found an affordable cornerstone for his plantation in 1986. It was an abandoned, cut-over 100-acre farm with dilapidated barns and shacks that had been built nearly a century earlier. Here is how Robert J. Smith described it in an article for the Center for Private Conservation (www.privateconservation.org): "Where once stately, open, pine-savannah forests had grown, there were now dense thickets of short, stunted, crowded, bent and twisted pine, gum and maple." In an interview with Smith, Burris recalled: "All kinds of trash was lying around. It was just wiped out. The barns on the land had been left to decay, as were the few trees that remained. The whole place was a total disaster—but it was affordable."

From that humble beginning arose a vision of what could be. The Burrises drew inspiration from Proverbs 29:18—"Where there is no vision, the people perish"—in writing their own manifesto for the land. "Our vision is to develop an ordinary piece of land and, with a plan and a commitment to lots of hard work, create a tree farm that will serve as a model to other tree farmers." That vision rested on five principles:

- Restoration of the land, buildings and forest.
- Conservation practices for trees, their primary crop.
- Preservation of the native live oaks, wildflowers and non-game animal species.
- Education through demonstrating the latest tree-farming practices to their neighbors.

- Perpetuation of the forests for multiple uses, including recreation, to "ensure a sustainable forest which will be sustaining for generation after generation."

Skeet, Gail and their four sons signed and framed the completed document, which hangs on the wall of their plantation home. Then, the "hard work" part began.

The first order of business was restoring and upgrading the tumbled-down buildings into serviceable barns and a remodeled cabin. In their first year, the Burrises also cleared wildlife food plots out of the thickets and planted their first trees. All the while, curious neighbors were watching—and their respect for the Burrises grew. Slowly, the family began to pick up other parcels connected with or near their property. Through 24 individual acquisitions, Cypress Bay grew to 1,038 acres, with another 2,000 acres leased from surrounding lands in order to carry out a wildlife management plan for native white-tailed deer, wild turkey and northern bobwhite.

Ground was prepared to start replanting the forests through bushhogging, thinning, burning, herbicide treatments and disking. Over time, some 200,000 trees have been planted at Cypress Bay, mostly longleaf and loblolly pines, the dominant trees of the southeastern forest ecosystems and marketable as timber. Because the plantation is also being used for hunting, hundreds of oak trees of five species were planted, including 449 fast-growing sawtooth oaks. In addition, scores of ornamental and fruiting trees (such as wild plum and wild pear) have been planted for aesthetic reasons and for game and non-game species.

Eradicated were the weedy gums and short maples that so often overrun cut-over or abandoned lands. Preserved were the native live oaks, some of which were giants covered with Spanish moss. If this meant cutting out competing trees, brush thickets and vines to "release" the majestic oaks, the Burrises did so.

There are no waterways running through Cypress Bay, so the Burrises built about fifty acres of ponds to enhance wildlife diversity and to make the setting much more attractive. Fish ponds, green tree reservoirs and duck ponds are seasonally planted and flooded. About 4,400 bald cypress trees have been planted around the ponds, which provide habitat for nesting ducks in the spring and summer as well as feeding and roosting habitat for migratory waterfowl that arrive in the fall and spend the winter. Cypress Bay's visitors include great blue heron, all-white great egrets, snowy egrets and wood storks. The storks are listed as endangered under the federal Endangered Species Act. A few unexpected visitors have staked their claims, too, such as beaver and American alligator. It wasn't exactly as planned, but it all goes to show how readily wildlife can adapt to newly created habitats.

Picture all this amidst a variety of food plots, firebreaks and wildlife corridors, as well as a mix of productive grasses, shrubs, grains, corn and sunflowers to feed the game and non-game species. The "vision" begins to emerge.

But there's more. The Burrises did much of this on their own, but far from all. In addition to enlisting the help of friends and neighbors, they coordinated their efforts with a number of private wildlife associations. As part of the South Carolina Waterfowl Association's Wood Duck Project, they erected 34 nest boxes that produce about 200 ducklings per year. In cooperation with the association's Mallard Project, Skeet has released about 1,400 ducks. Son Charlie and a Hampton County Boy Scout troop have placed about 20 bluebird boxes that have produced about 100 young eastern bluebirds. Nest boxes have also provided homes for hooded mergansers, eastern screech owls, great crested flycatchers, Carolina wrens, warblers and purple martins.

The Burrises work with Quail Unlimited, the Quality Deer Management Association, the National Wild Turkey Federation and Ducks Unlimited—and it has paid off for wildlife and for the enterprise. They're getting top dollar for duck hunting leases, upland bird (turkey and quail) leases and duck blind leases. "(Skeet's) wildlife management practices have been so outstanding and are producing such high-quality hunting that people are queuing up across the Southeast for the opportunity to obtain a lease," wrote Robert J. Smith of the Center for Private Conservation.

Fifteen years of work has been followed with an assortment of honors, including South Carolina Tree Farmer of the Year, Southern Regional Tree Farmer of the Year (twice) and, most recently, National Tree Farmer of the Year. In April 2001, the Burrises were named Private Conservationists of the Year by the Center for Private Conservation in Washington, D.C.

And yet, the work at Cypress Bay is never done. The habitats, as one should hope, are becoming increasingly rich for wildlife, and not just huntable species.

"Our vision is getting larger and larger," Skeet Burris said, "and I'm still busting my tail every second I get. But the real credit goes to a lot of people who gave us their ideas. We had input from a lot of folks. I've simply utilized other landowners' good ideas."

Cypress Bay is a model for private conservation—and also for participatory environmentalism, as the work there involved nonprofit and public interests as well as the Burris family. In addition, it is an outstanding example of multiple-use stewardship, the kind that Aldo Leopold accomplished two generations ago in Sauk County, Wisconsin, but which he never lived to see spread across the nation.

Profit and stewardship not only can coexist, but should. As Leopold recognized decades ago, the recreation market benefits the sportsman and the environment as much as it does the landowner. As pessimistic as he was at times, Leopold never gave up on the power of market forces. There was no reason to reject them, he argued, simply because "such tools are impure and unholy" in the eyes of some. At Cypress Bay Plantation, those free-market tools have rebuilt a forgotten patch of land into a small Eden.

11
Back to School

"Earth Force" and the Council for Environmental
Education help America's young people get their
hands around stewardship

Much of what passes for environmental education in the United States is strong on advocacy and alarmism but weak on science and practical application. However, some national programs make the grade.

ENVIRONMENTAL EDUCATION IN AMERICA'S public schools gets a "C" for political correctness but a "D" for indoctrinating kids with a doomsday philosophy that's straight from the eighteenth-century textbooks of Thomas Malthus. James M. Taylor, the managing editor of *Environment and Climate News*, experienced this firsthand when he accompanied his children on an elementary school nature walk. Taylor wrote how he listened in surprise as the guide claimed that Florida's 2001 drought happened because "too many people cut down trees to make houses" and that "the worst thing that ever happened to Florida was the invention of pesticides and air conditioning."

The notion that people are nature's enemy, rather than part of the answer for sustaining it, is a recurring message in the nation's schools. Apocalyptic textbooks and a federal education effort run out of the EPA's Office of Environmental Education threaten to produce a generation of kids who are too scared to manage their future. A recent study of more than three hundred environmental books and guides used in the nation's public schools revealed some shocking—and largely unsupported—"facts" that are being taught in the classroom. For instance, did you know?

- Earth's natural resources will "become so depleted that our very existence will become economically and environmentally impossible."
- World petroleum supplies "will only last another year or so."
- If global warming continues, "New York City would almost be covered with water. Only the tops of very tall buildings will be above the water."
- Judeo-Christian philosophy and "the rise of capitalism" are among the reasons for today's environmental problems.

These examples come from educational materials reviewed by Michael Sanera, head of the Center for Environmental Education Research at the Claremont Institute in California.

To counter such misinformation, Sanera and Jane Shaw wrote *Facts, Not Fear: A Parent's Guide to Teaching Children about the Environment*. Published in 1996, the book sorted out commonly accepted facts about the environment from the many exaggerations and outright falsehoods in contemporary science textbooks and mass media reporting. It is the first guidebook to help parents counter irresponsible claims by powerful environmental extremists, to separate myths from scientific realities, speculation and theory from proven fact, and to answer children's frequently asked questions about the environment. Further work along these lines has been sponsored by the Environmental Literacy Council. A number of nature centers, including the exemplary Riveredge Nature Center near the headwaters of the Milwaukee River in Wisconsin, draw upon the fundamentals of the scientific approach to select sound information to include in their educational work.

In welcome contrast to the defeatism and one-sided advocacy evident in so much of what is taught to the nation's youth are programs such as "Earth Force," which has won three "Points of Light Foundation" awards since its inception in 1994, and the Council for Environmental Education, which has offered young people a civic-minded approach for thirty-five years.

Funded by the Pew Charitable Trusts and aimed at middle school students, Earth Force taps into two converging national trends: the desire of young people to act in behalf of the environment and their willingness to help their communities. Earth Force provides coast-to-coast examples of students practicing hands-on environmentalism close to home—and learning alternatives to mere acceptance of top-down control.

Educators across the United States turn to Earth Force (www.earthforce.org) for innovative ways to engage young people through two basic programs: Community Action and Problem Solving (CAPS), and the Global Rivers Environmental Education Network (GREEN). CAPS combines the best practices of environmental education, civic engagement and service learning, while GREEN helps young people protect rivers, streams and other waterways by merging hands-on, scientific learning with civic action. Earth Force has a fifteen-member national youth advisory board and eight site offices nationwide, and distributes resource guides and "bike action packs" to teachers and students.

"We're really about a process, not about issues," said Kris Maccubbin, communications director for Earth Force. "We're about kids doing balanced research to come up with effective solutions to work within the system and to build coalitions. We work with educators, but we tell them this is not about advancing a particular environmental agenda."

Earth Force students follow this six-step process, which is designed to prevent them from taking action until they've done their research.

1. Checking it out. Ask yourself: What strengths and problems or threats are present in our community?
2. Deciding what's wrong. Ask yourself: What problem or threat do we want to work on?
3. Sleuthing. Ask yourself: What are people already doing about this problem or threat? What laws, policies and practices are currently in place? Who makes or enforces the policies or laws, and how effectively? What are the different views on causes of the problem or threat and what should be done about it? What changes do we want to see in policies or practices?
4. Deciding what to do. Ask yourself: What can we do to bring

about the change we want to see? What are our choices for making a difference with this problem or threat?

5. Taking action. Ask yourself: What steps will we take to carry out our strategy?

6. Looking back and ahead. Ask yourself: How did everything go? What would we do differently? What can be done next?

Programs director Vince Meldrum said that Earth Force encourages young people to venture out into their communities, to talk with people on all sides of a problem, and not to act on emotional first instinct. "Part of what kids try to determine is what are the costs and benefits for each action. Just because something is environmentally sound, it may not be the best option available for that community. We want them to come up with solutions that meet as many community needs as possible."

Maccubbin said that Earth Force participants occasionally "play interesting roles as community catalysts," getting the local Sierra Club chapter and the Chamber of Commerce to talk about stalemated problems. It's hard to say "no" to middle school students who have done their homework and don't understand why adults sometimes refuse to do the same.

In Charleston, South Carolina, Jeff Erickson's sixth-grade classes at Rivers Middle School embarked on a project to improve a decaying urban environment. They mapped about forty abandoned or run-down buildings in their neighborhood, learned that it's much easier for absentee landlords to pay fines than fix things, and encouraged city officials to adopt a mix of carrots and sticks to clean it up.

In Philadelphia, students at Shawmont Middle School wouldn't take regulation for an answer when the EPA shut down their school's water fountains because of high lead levels. Follow-up tests precipitated by the students proved that the EPA had over-reached on two fountains. The Earth Force students also put pressure on the EPA and school officials to quit passing the buck and come up with a permanent solution.

About 25,000 students take part in Earth Force nationwide, working on projects that range from habitat restoration to creating small ponds to improving the urban environment.

Programs run by the Council for Environmental Education (www.c-e-e.org) reach about 50,000 educators a year. Its mission is "to provide environmental education programs and services that promote stewardship of the environment and further the capacity of learners to make informed decisions." The council is a founding cosponsor of Project WILD, Project Learning Tree and Project WET; it administers Project WILD, Project WILD Aquatic and WET in the City at a national level. The WET in the City program, developed to reach underserved students, brings education on the stewardship of water resources to urban K-12 educators through community-based networks. With their commitment to balanced, unbiased instruction, these programs are among the most successful environmental education efforts in the United States.

Organizations such as Earth Force and the Council for Environmental Education are helping young people learn hands-on lessons about citizenship and stewardship that will last a lifetime. If only the same could be said for those American kids who aren't lucky enough to be getting a balanced environmental education.

12

Reservation Conservation

Arizona's White Mountain Apaches are managing their natural resources and turning a profit for themselves and the land

The White Mountain Apache tribe manages its natural resources and outdoor recreation programs as it would a business, an approach that is good for the land, the water, the wildlife and the tribe's 13,000 members. But it hasn't always been easy dealing with government agencies that fear losing control.

NO LESS A FISHERMAN THAN IZAAK WALTON, the seventeenth-century author of *The Compleat Angler,* described the loach minnow as "a most dainty fish: he breeds and feeds in little and clear swift brooks or rills, and lives there upon the gravel, and in the sharpest streams: he grows not to be above a finger long, and no thicker than is suitable to that length."

By the early 1990s in Arizona, the loach minnow apparently was so "dainty" as to be considered threatened by the U.S. Fish and Wildlife Service. The concerned agency dispatched a regional administrator from Phoenix to the White Mountain Apache Reservation, some two hundred miles to the northeast, to investigate reports that the loach was struggling to survive in a stream that ran through the tribal center of Whiteriver. The only way to save the loach minnow, the Fish and Wildlife emissary told shocked tribal leaders, was to move the entire town. The last time a federal agency had issued such an order, about a hundred years earlier, the White Mountain Apache had only recently laid down their arms after years of struggle against a white government intent on subjugating them.

Needless to say, the only person who moved quickly—on the

next car back to Phoenix—was the hapless Fish and Wildlife bureaucrat. The White Mountain Apache flatly refused to uproot a town of 2,500 people to save loach minnows that the tribe's conservation practices were already saving. Thanks to a lack of respect for the tribe's sovereignty on its 1.6 million acres, the Fish and Wildlife Service took years to earn its way back on the reservation. Today the agency has an office just outside the reservation's borders, and relations are improved. Even so, the tribe has stuck to its philosophy of natural resource management: Private conservation and local control produce tangible results.

"One-size-fits-all may work for bureaucrats in Washington, D.C., but it doesn't produce results on the ground," said Jon Cooley, the former director of the White Mountain tribe's wildlife and outdoor recreation division. "Incentives are what have worked for our tribe."

The majestic White Mountain reservation stands as proof of Cooley's statements. From the Sonoran desert at 2,500 feet to alpine elevations of 11,000 feet, its landscape ranges from oak chaparral to mixed conifer forests. The reservation holds one-third of Arizona's cold-water reserves in the form of lakes, rivers and streams. That includes twenty-four man-made lakes stocked with trout and other fish. The White Mountain reservation also features whitewater rafting and boating areas and wild canyons, making it a "must-visit" destination for thousands of fishermen, campers, hikers, boaters and other tourists.

For hunters, the White Mountain reservation is nothing short of paradise—they are bagging trophy elk in envious numbers. The tribe has been managing its elk herd on its own since the late 1970s, overcoming legal and bureaucratic hurdles along the way. Its wildlife management practices have produced one of the highest-quality elk herds in North America.

"Entrepreneurship by the White Mountain Apache tribe changed the quality of elk hunting dramatically," wrote Terry L. Anderson and Donald R. Leal in *Enviro-Capitalists: Doing Good While Doing Well.* "Under tribal management, the emphasis has been on greatly reducing hunting pressure on immature bulls so they will have a chance to grow to record size." Hunters pay as much as $12,000 for a seven-day guided hunt in search of a record-book bull elk. Tribal revenues from elk hunting alone have climbed to about $1 million a year, and much of the money flows back into improving the habitat.

Some concessions have been made. For example, livestock grazing has been curtailed in some areas to give the elk more forage. Tribal biologists review timber sales to make sure that cutting doesn't hurt the elk herd. No logging is allowed in high-country areas, riparian zones or mountain meadows. In those areas where logging is allowed, it's staged to avoid calving periods, and roads are closed after logging is finished to minimize disruption of elk habitat.

Fires that raced through the White Mountain area in the summer of 2002 and limited fires on subsequent occasions hit hard at the tribe's logging business, destroying what was initially estimated at $200 million worth of timber. In some areas, it may take a century or more for the damage to repair itself. But the tribe isn't waiting for time to solve its problems. Salvage operations are harvesting fire-damaged timber that would otherwise go to waste, and sales of that timber are providing much-needed cash. The long-term challenge is to ensure that logging and habitat protection continue to coexist.

"From a resource management standpoint, the tribe is going to have its hands full for a while," said Cooley, who was executive director of the Southwest Tribal Fisheries Commission and is now in wildlife service with the State of Arizona. "As the landscape heals over time, however, I could see how [the fire] might have benefits. For some species, it actually will improve the habitat. For others, it will not."

The fire caused the most damage on the east side of the reservation, away from the major recreational areas to the west. It burned through ponderosa and rugged canyons alike, destroying habitats for many species and raising the potential for erosion and even floods. Runoff from denuded canyons will cause sediment problems in streams and lakes, which will affect fisheries.

The fire hurt tourism, hunting and fishing in 2002, probably far more than it had to. "It's been business as usual in a lot of ways," Cooley said, and the tribe is spreading the word that most areas were untouched by the blaze.

Hunters on the White Mountain reservation may buy permits for bear, javelina, wild turkeys, quail, tree squirrels and cottontail rabbits. Anglers can fish for native Apache trout (catch-and-release in streams only), rainbow, brook and brown trout, arctic grayling, bass and northern pike. The tribe also offers a "rent-a-lake" program that offers exclusive use of Cyclone or Hurricane lakes for a day or a weekend. For hunting, fishing, camping and other fees, visit (www.wmatoutdoors.com).

Fee-based recreation has proven to be a blessing for a young tribe (the average age in 2002 was twenty-one) that needs an independent, sustainable economy. About $2 million in annual revenues are spent on habitat enhancements, protecting plant species that are important to the tribe for cultural reasons, improving recreational facilities and providing seed money for business startups elsewhere on the reservation. The program has also generated jobs for hunting guides, outfitters, cooks and more. "To me, resource management is every bit as much of a business as running a factory," Cooley said.

Challenges still remain for the White Mountain Apache and their hands-on approach to protecting their own environment. The U.S. Fish and Wildlife Service isn't keen on the tribe maintaining hatcheries for trout that aren't native to east-central Arizona. But the revenues from those trout are essential to the tribe's protection of the rare Apache trout. No fishing revenues means no rare species protection, Cooley explained. "All of our work here, from protecting watersheds to sensitive species to erosion control to our eco-team planning, revolves around maintaining and growing our revenues from fee-based recreation. We're trying to take a more long-term pragmatic approach and move away from the perpetual crisis mode. An incentive-based system helps us do that, where a command-and-control model doesn't."

The tribe's incentive-based approach to stewardship of natural resources has carried over to how it protects "heritage resources," such as cultural, historical and archaeological assets. A heritage program launched by the tribe in the early 1990s has more or less run parallel to the larger natural resources plan, reflecting the connection between people and the land.

The White Mountain Apache model has evolved over time; it

will have to adapt to circumstances like the extensive fires early in the twenty-first century; and in many ways it is unique to that tribe. The basic principles, however, are universal: Give people the ability to do well by doing good, and the people and the land will profit. On Arizona's magnificent White Mountain reservation, twenty-first-century independence means charting a course that is good for both.

13

Bravo!

How a Wisconsin utility initiated a partnership to protect rainforest and produce ecosystem services in Belize

Wisconsin Energy doesn't sell electric power south of Beloit, much less in Belize. But the lack of any direct business interest didn't stop the company from invest-ing $2 million in a sustainable forest project to help Central American land, wildlife and people. What's more, the Rio Bravo project is capturing tons of greenhouse gases and serving as an example to others.

DICK ABDOO DOESN'T FIT THE PROFILE of someone who would spend a lot of time worrying about the environment. He was a longtime chairman of Wisconsin Energy, an $8.4 billion, Milwaukee-based company that routinely does something many environmentalists hate: It burns coal to generate electricity. But Wisconsin Energy also did something else that goes far beyond any antipollution rules set by federal and state regulators. Abdoo's company is helping save a rainforest in Belize.

Belize is a small Central American nation south of Mexico's Yucatan Peninsula, east of Guatemala and north of Honduras along the Caribbean coast. In Belize's northwest corner lies the 260,000-acre Rio Bravo Conservation and Management Area, a part of the culturally and biologically significant Mayan forest region. It's a wild place populated by endangered species such as the black howler monkey, the jaguar, a variety of migratory and resident birds, and some of the densest mahogany stands on earth. Rio Bravo contains forest cover types protected nowhere else in Belize.

Nearly ten years ago, much of this unique area was being threatened. Native farmers were cutting down mahogany trees at an alarming rate, taking down much of the forest canopy with them. For the people who lived there, it was a matter of making a living. Saving habitats for howler monkeys or cleansing the world's atmosphere of carbon was understandably the last thing on their minds.

Enter Dick Abdoo. A former member of the Nature Conservancy's national board of governors, he learned of the disturbing losses in Rio Bravo and began to talk with scientists from other organizations that might have a stake in providing incentives for a sustainable forest project. Programme for Belize was already working to protect land; Wisconsin Energy, the Nature Conservancy and others soon joined in the effort.

About that time, the provisions of the 1990 Clean Air Act were kicking in, and Abdoo engaged the U.S. energy secretary, Hazel O'Leary, in talks about Wisconsin Energy's plans to buy only clean coal from mines that captured methane and took other environmentally sensible steps, and to begin experimenting with carbon sequestration. Former Wisconsin governor Anthony Earl, who had once been the state's secretary of natural resources, pitched in. So did the Nature Conservancy's Tia Nelson; daughter of Earth Day founder Gaylord Nelson.

"We set out to show that with a risk-informed, incentive-based approach, we would develop a sustainable forest management project that would allow native farmers to harvest as much wood as they needed without harming the ecosystem," Abdoo recalled. "We believed that, even in a poor country, forests can be better managed, wildlife and plant species can prosper, and people can be better off economically."

What made the project different was the coordinated effort to use better forest management to capture carbon—and thus reduce greenhouse gases that most scientists believe are aggravating a global warming phenomenon. Scientists agree that forest conservation and management play an important short-term role in global climate patterns by literally absorbing atmospheric carbon dioxide. Forests breathe in carbon dioxide and exhale oxygen, which is among the many reasons why deforestation in one part of the world poses a threat elsewhere. Hardwood forests are best at soaking up carbon.

The Belize partners invested a total of $5.6 million to fund the first ten years of a forty-year carbon sequestration project. It was built around the purchase of endangered, moist, subtropical broadleaf forest lands to supplement the holdings of Programme for Belize. The group's money also helped in setting up a sustainable forestry program, hiring staff, and providing security to guard against theft, misuse and fire.

The project's benefits have been multiple and measurable. Sustainable forest management in Rio Bravo has:

- Reduced soil erosion and pollution of surface waters.
- Maintained greater biodiversity.
- Promoted a more sustainable local economy and created new jobs.
- Incorporated new concepts, such as global climate change, carbon sequestration and sustainable development, in local curricula. This will help sustain change over generations.
- Maintained farming productivity.
- Transferred knowledge about sustainable forest practices to other private lands in the region.

Scientifically verifiable carbon sequestration activities take place on about half of the Rio Bravo's 260,000 acres. It is expected that the project will reduce, avoid or mitigate about 2.4 million metric tons of carbon over forty years. Third-party inspections verify the greenhouse gas reductions, and the project partners report each year to the U.S. Initiative on Joint Implementation (USIJI). The USIJI program encourages American entities—such as Wisconsin Energy and the Nature Conservancy—to invest in innovative greenhouse gas reduction projects in developing nations. It promotes technology transfer, sustainable development and science that can be used to chart what does and doesn't work.

The Rio Bravo project was one of the first so-called "joint implementation" projects of its kind in the world to be fully funded. It now serves as a model for similar projects, including American Electric Power's carbon sequestration project in Bolivia and projects in Brazil that count the Nature Conservancy among its partners. (By the way, not all U.S. utilities are getting high marks for their environmental programs in Central and South America. Duke Energy International, a subsidiary of the North

Carolina–based utility, pulled out of the proposed Chalillo dam and reservoir project in Belize after getting some 20,000 letters and e-mails from activists who claimed that the project would ruin a unique stretch of the Macal River.)

"All of a sudden, carbon sequestration has hit the radar screen of the environmental agencies and the environmental groups, which didn't seem to want to grasp this at first. Now, they're talking about it," Abdoo said.

Abdoo had hoped the federal government would recognize the work being done in Rio Bravo and elsewhere with a voluntary carbon-offset trading program for utilities. He believes that flexible, market-based emission-reduction programs can be encouraged if Rio Bravo continues to work as planned and others follow. So far, however, the feds haven't budged—even though market-based systems for trading greenhouse gas emissions are emerging elsewhere.

Under this incentive-based approach, pollution allowances can be bought and sold freely on the market. Developing nations can hold their allowances for industrial development or sell them for revenue to invest. Companies in developed nations, such as Wisconsin Energy in the United States, can buy allowances to continue their growth, but at a price that will encourage them to reduce emissions. Existing emissions trading organizations include the Chicago Climate Exchange (www.chicagoclimatex.com/), which has signed up twenty-five midwestern companies; the Greenhouse Emissions Management Consortium (www.gemco.org), which works with eleven Canadian companies and has brokered a deal involving Iowa farmers; CO2e.com (www.co2e.com) in New York, which facilitates twenty-four-hour online trading of emissions reductions; KEFI-Exchange (www.kefi-exchange.com), which is based in Alberta, Canada, and completed the world's first online trade in 1999; International Emissions Trading Association (www.ieta.org), a London-based nonprofit; Climate Change Central, another Canadian enterprise (www.climatechangecentral.com); and the Emissions Marketing Association (www.emissions.org), an American networking group.

While it has been frustrating for Abdoo that Rio Bravo has yet to result in tradeoffs or credits, he did not back off the company's work in Belize. He brought a skeptical Wisconsin Energy board of directors to Belize to review progress, and he sponsors exchange programs that have involved nearly one hundred high school students and science teachers since 1995.

The Rio Bravo didn't directly add a penny to Wisconsin Energy's bottom line, but Abdoo saw other rewards that are no less tangible. "Science is gaining valuable information about carbon sequestration, the people and land in Belize are better off, and the world is a better place because more carbon is being taken out of the land," he said. Plus, Wisconsin Energy is learning that doing good for the environment helps the company do well financially in surprising ways. "The more you can reduce environmental damage relative to your competitors, it becomes a competitive advantage in today's world. Best of all, we're not doing this because of command-and-control rules. We're doing this because it has become a basic part of our business plan. We're releasing American ingenuity."

Someday soon, those who believe in command-and-control methods may recognize that it's better to set goals and provide incentives than to prescribe cleanup techniques that may lag far behind the latest technology and science. Wisconsin and Belize are separated by thousands of miles, but the lessons being learned through the Rio Bravo project may wind up being exported back home to a country that could use new solutions.

14

It Takes a Village to Raise a Rhino

How community-based conservation and
cooperative neighbors are saving rhinos,
elephants and more in Africa

*There are few continents where the biotic diversity is richer—and the threats to
its existence more severe. Warfare, corruption, poverty, disease and collectivism
have hampered conservation in Africa, but "hands-on" environmentalism is working where neighbors collaborate and communities democratically manage their
own resources.*

AS RECENTLY AS 1994, THE OBITUARIES FOR Africa's black rhinoceros
were being written in the Western press. A front-page story in the
Los Angeles Times called efforts to save the black rhino "a losing
battle" and predicted, "by all accounts, [the black rhinoceros] may
be doomed." Headlines that same year in *Buzzworm* magazine
were nothing short of grim: "The Rhino Chainsaw Massacre: Why
Rhinos Will Not Survive the Century."

What the worried and somewhat apocalyptic press didn't
know in 1994 was that efforts begun four years earlier in Zimbabwe's lowveld (Afrikaans for "lower plains") conservancies were
already turning the corner in the fight to save Africa's black rhino
from extinction. The Zimbabwe experiment and others somewhat
similar to it in Zambia, Namibia, Botswana and South Africa were
replacing the command-and-control model that had dominated
African conservation since colonial times with a community-appro-
priate approach based on property rights. By treating wildlife as a
marketable commodity rather than a resource off-limits to people
who must endure the nuisance and get none of the benefits, Zimbabwe's four lowveld conservancies (Save Valley, Bubiana, Chiredzi

River and Midlands) had by 2000 dramatically reversed the decline of the black rhino as well as other wildlife and plant species.

Since 2000, a form of ecoterrorism has eroded some of those gains. Economic collapse and social disarray brought on by state-sponsored land redistribution and corruption have encouraged squatters and paramilitary forces to seize farms, game preserves and parks. By the fall of 2003, the *New York Times* reported, as many as two-thirds of the animals on Zimbabwe's game farms and wildlife conservancies had been wiped out.

Described as an "unholy slaughter" by one Zimbabwe conservationist, this decimation of wildlife has been far worse in the parks than in those private conservancies that have managed to hold together. And in other parts of Africa, where totalitarian thugs are not in charge, the conservancies are prosperous places for animals and people alike. A *New York Times* article in 2004 pointed out the remarkable maintenance of game, including the imperiled Grevy's zebra, in Kenyan conservancies while surrounding lands were stripped of grass, bush and game. At Lewa Conservancy the conservationist owners reported: "In 1977 there were 81 Grevy's on Lewa. Today, when they are in a rapid decline elsewhere, they number 400. This figure represents almost 17% of the world's population—one of only three groups under protection, and the only one managed privately."

To understand the evolving story of southern Africa's conservancies and how dramatically they have departed from the typical African model of wildlife conservation, one must first go back in time to the colonial era and examine some of the roots of a paternalistic and disastrous conservation ethic that plagues the continent to this day.

Before white Europeans arrived, many Africans followed natural resource use traditions that had worked for generations. (Of course, exploitation of resources was often limited by the tools available.) There were taboos against eating certain animals; sacred forests of spiritual importance were preserved; hunting was restricted in certain areas and the killing of pregnant animals was prohibited. In many cases, rules were adopted by subsistence hunters with the hope that the resources they relied upon were not going to be overused.

With the use of guns came overhunting in the "open range"—often encouraged by colonial governments for development purposes. But it became so severe that even a few colonial authorities began to worry. Rules that at first were intended to stop white hunters from shooting everything in sight were soon applied as well to native people, who much more than the white Europeans were killing wildlife for food.

"Laws against hunting by natives...precluded the exploitation of what had been for literally millions of years a critical source of food for African communities," said Fred Nelson, a Tanzania-based consultant to the Sand County Foundation. "Placing wildlife off-limits to Africans essentially acted to alienate the resource from them, meaning they no longer had any interest in the fate of the wildlife since their traditional subsistence ties to wild animals and utilization practices had become criminalized.... Rights to use wildlife locally were largely eliminated, and thus whatever incentives for sustainable exploitation had existed were extinguished."

Next came the creation of "protected areas" and reserves that fenced people out and animals in, but only for the enjoyment of the few. At the same time, massive state-sponsored slaughters of game were under way to open up land for agriculture and livestock and to reduce the disease threat.

"Initially," Nelson wrote, "indigenous people were often allowed to remain in reserves, but eventually it became widely accepted that people and wildlife must be separate for conservation to succeed, and natives were evicted from their traditional homes and resettled outside of these new protected areas. Relocations of people were the rule rather than the exception in creating national parks and exclusive reserves. Colonial conservation efforts thus deprived people not only of wild animal resources, but also of vast tracts of land that had formed the basis of livelihoods for generations."

Thus, a perverse preservation ethic that was neither biologically nor politically sustainable took hold. To native Africans, effectively segregated from a land that had been their own, the colonial notion of "game conservation" came to mean that wild animals could stray from reserves, trample crops, kill livestock and menace humans without penalty—and that self-preservation efforts would be dealt with severely. It is an attitude and a reality that exist to this day, primarily because the protectionist mentality of reliance on "preserves" and segregating people from wildlife didn't change when the colonial era ended.

David Hulme and Marshall Murphree captured that mood when they interviewed a villager in Uganda for their book, *African Wildlife and Livelihoods: The Promise and Performance of Community Conservation.* "It is not fair," the villager said. "If their animals [the National Park] come on our land and do a lot of damage, we get no compensation. If our cattle stray on to their land, we are punished."

While foreigners may relish the opportunity to see a bull elephant or a lion in the wild—or even to watch such creatures on television—rural Africans have been conditioned for generations to see wildlife as a nuisance to be exploited or even eradicated rather than a treasure to be preserved. Wildlife to them is either something to eat, something that will destroy your crops or livestock, or something that might even hunt and kill your children. This view is a natural consequence of public ownership gone wild: When everyone owns something, benefits seem distant and the costs become a drag on daily life.

"The parks have no value to the people. In fact, they have a lot of costs," said Brian Child, a fifth-generation African who has been program director for Zambia's South Luangwa Area Management Units (SLAMU). (This sprawling, community-based conservation project is funded in part by the Norwegian government.) "The elephants go out of the parks and eat the crops. The lions go out of the parks, kill the livestock, kill the cattle, even the people. There's a global benefit [to the wildlife], but there's a local cost. The poor people who are here—and they have absolutely nothing, other than a couple of pots and a hut—are paying for global biodiversity. I don't think that's socially or politically sustainable."

In modern Africa, the colonial conservation model of public ownership isn't working for yet another reason: Laws are not being enforced, which can lead to a "tragedy of the commons" in those areas where poaching is endemic. "If wildlife is to be conserved through existing legal prohibitions on use and management of protected areas, then law enforcement must play a primary role in executing those policies," Fred Nelson wrote. Instead, many nations don't have the money to hire wardens to drive away the poachers, which means the worst of all worlds for the unprotected—and unwanted—wildlife. A law that is unenforceable can be worse than no law at all.

Since the 1980s, observers of the conservation disaster unfolding in Africa have consistently criticized central ownership of resources.

• Zimbabwe native Graham Child, in his 1995 book, *Wildlife and People: The Zimbabwean Success,* suggested that restrictive wildlife utilization policies and central management systems have done more to destroy Africa's wildlife than any single factor with the exception of habitat loss.

• Rob Barnett, author of *Food for Thought: The Utilization of Wild Meat in East and Southern Africa,* wrote that the growing consumption of bushmeat is a result of wildlife being viewed as a free-for-all resource. There's a lack of public enforcement on the one hand, he wrote, and an absence of local resource tenure and utilization rights on the other.

• George Ayittey, author of *Africa in Chaos,* concluded that unworkable centralized management systems have generally been a biological and socioeconomic disaster for Africa, just as socialist experiments have produced calamitous results in rural areas.

• Michael De Alessi, former director of the Center for Private Conservation, had this to say after his visits to the Save Valley and Bubiana conservancies in 1997: "When poverty is widespread, people worry about where their next meal is coming from. For them, protecting wildlife for its own sake is an unimaginable luxury. If wildlife does not pay, it will be replaced by something that does," such as crops and livestock.

All these factors contributed to the decline of the black rhinoceros (*Diceros bicornis*), a savannah and thicket species that feeds on young twigs, leaves and shoots. The "black rhino" is not really black, but was probably so named to distinguish it from the more common "white rhino" (*Ceratotherium simum*), which has a wider lip; the Afrikaans word for "wide" sounds much like "white," and thus the familiar name. In fact, both are actually gray, but the distinction based on a misunderstanding nonetheless stuck.

The black rhino is an herbivore and a basically shy one at that, although the World Wildlife Fund correctly describes these 1,400-kilogram animals as "hostile when disturbed." Dangerous as they may be, black rhinos became a target for poachers because of the demand for rhino horn, mostly in the Arab world. In Yemen and elsewhere on the Arabian peninsula, the horns were used to make dagger handles—for any one of which tens of thousands of dollars would change hands. For a dagger handle, a black rhino would be killed and left, hornless and no longer capable of producing tangible benefits for any people anywhere.

In 1970, there were about 65,000 black rhinos in Africa. By 1995, the number was down to about 2,500, in part because of habitat loss but also due to government policies that made a precarious situation worse. Poaching increased throughout the 1980s, despite a ban on rhino horn trading, dehorning programs and stepped-up enforcement efforts. "As late as 1994, poachers continued to kill rhinos despite radio collars, dehorning of hundreds of rhinos, use of heavily protected animal sanctuaries, and a shoot-to-kill policy that left 178 suspected poachers and four game wardens dead," wrote Michael De Alessi in his 2000 report for the Center for Private Conservation.

The future of the black rhino finally took a turn for the better in the late 1980s when the Zimbabwe Department of National Parks and Wildlife Management began moving rhinos out of the Zambezi Valley to private lands with owners willing to nurture the beleaguered black rhino back to healthy populations. Ownership of these majestic animals remained with the national government, unlike the effective ownership and utilization authority that landholders had for most other species. By 1991, a British-based organization with a history of work in Zimbabwe, the Beit Trust, provided some seed money to initiate the experiment. But there was a very real problem to overcome: convincing ranchers to welcome wildlife back to land that had been altered (and not for the good) by a century or more of grazing and agriculture.

Mother Nature decided to help in a strange and initially unwelcome way. A severe drought in the early 1990s left many ranchers on the ropes. The cattle herds were devastated, but the wildlife endured. Ranchers began calculating the advantages of switching from cattle to both consumptive and nonconsumptive uses of wildlife. It was right for the animals—and right for the landowners. Sticking with cattle would yield profits in the range of 1 to 2 percent, members of the Save Conservancy calculated, while wildlife management would bring a return on capital of 10 percent or more. "When the drought ended, we decided never to return to beef, which had been running at a loss for years," said Derek Henning, a Save member, to De Alessi.

A remarkable transformation in the Save Valley was under way, thanks to private initiative and a partnership that eventually grew to more than twenty owner-members and about thirty properties spread over nearly 900,000 acres in southeastern Zimbabwe. The owners formed a "private conservancy"—which is today Africa's largest—although doing so took some creative thinking.

The story was much the same in the Bubiana Conservancy, which at one time had seven ranchers, ten properties and 400,000 acres. Cattle ranching in the Bubiana Conservancy was not eliminated, but it became a secondary source of income.

"In the cattle heyday, farmers viewed the wildlife as competitors to the cattle and eradicated [the wildlife]," said Ken Drummond, a member of the Bubiana Conservancy. "And the cattle, in the numbers they were running in, eradicated the habitat to a degree that the wildlife couldn't survive anymore. Nature was kicked out of the way to make room for agriculture."

Fragile soils became depleted and native plants and grasses became harder and harder to find. Beginning in the 1990s, however, the ranchers in the Bubiana region decided it was time to form a conservancy and quit trying to change what nature intended. "The right horse for the course here was nature's animal, not a domestic animal," Drummond said.

What is a conservancy? A study conducted by Price Waterhouse for the Save, Bubiana and Chiredzi River conservancies arrived at this definition:

> The term conservancy can be applied to any number of properties which are amalgamated into a single complex in order to enable more effective management, utilization and protection of some or all natural resources in that area. In the case of the

lowveld Conservancies, they are developed on the principle that in arid regions, rangeland resources need to be managed at a larger scale than individual farms, in order to cope with a variable and unpredictable environment. The Conservancies under consideration are all managed in terms of agreements between the members, although the content of these agreements varies according to the aims of the members. The main focus of the conservancy agreements is cooperative management of the wildlife resource. However, this focus can also be extended to cooperative business ventures between conservancy members.

Simply put, wildlife requires room to roam, feed and breed, and the private landowners in the Save Valley, Bubiana and Chiredzi River conservancies could produce more of that necessary room to roam, feed and breed by collaborating as neighbors.

"Individual development of wildlife farms would mean the fencing off of each property," explained Guy Barber, owner of the Barberton Lodge, which overlooks the Bubiana Conservancy from atop a *kopje* (hill) that rises above the savannah. "You have a much more flexible and far improved wildlife environment by doing it collectively. It's more profitable than cattle because it's more sustainable. And it withstands the elements far better than does conventional agriculture."

Ken Drummond was among those who helped set a new course in the Bubiana region. He spoke with other ranchers about the necessity to change and to look beyond some traditional ways of using private property. "I was able to persuade them that if you've got enough room to roam for animals, you've got a wildlife business. If you don't have enough room, you've got a zoo," he recalled during the filming of *Conservation Pathfinders*, an NETA/PBS documentary. "The conventional game farming approach is very short-sighted. It's just one step better than nothing. The right thing to do if you're talking about farming nature is to let nature happen, and for nature to happen, you need space."

The commercial value of wildlife in Africa is captured in several ways. First, there are "consumptive" uses such as trophy hunting. Typical "nonconsumptive" activities are bird watching, photo safaris and other opportunities to see Africa's "Big Five" species—rhino, elephant, lion, leopard and buffalo—in the wild. A generation ago, it was difficult to find any of these species in abundance on the conservancies. Today, thanks to habitat restoration, around-the-clock poaching enforcement and the reinvestment of wildlife-related profits into ecotourism, it's possible to see the Big Five plus zebra, giraffe, warthog, wildebeest, hippopotamus, cheetah, impala, eland, bushbuck, sable, hundreds of bird species and more.

At the Save Valley Conservancy, one of the most treasured experiences is tracking and glimpsing one of the black rhinos. Thanks to the conservancy's efforts, it's now possible to do so—if you don't mind the long walk and you have an ample supply of patience.

Black rhino numbers on the conservancies have grown at a rate that has exceeded expectations: 35 rhinos had been relocated to the Save conservancy by 1193; there were 57 by 1997, and 74 at the end of 2000. With further population growth, plus some recent translocation, numbers at Save are above 120, despite losses of a few young to poachers' snares and a few adults to poachers' bullets. Bubiana started with 38 introduced rhinos in 1991 and had 69 by 1998 and 80 by the end of 2000. The black rhino population growth rate in the Chiredzi River and Midlands conservancies has been similar.

In the first seven years of the conservancies, not a single rhino was poached and black rhino numbers were growing in Zimbabwe for the first time in decades. Unfortunately, that changed in 2001 when five black rhinos were killed in Bubiana and two in Save, the byproduct of political unrest. In late 2003, seventeen black rhinos were moved from Bubiana to the Bubye River Conservancy by private conservation groups worried about the animals' fate. Others were dehorned to prevent poaching. The best estimates were that close to one hundred black rhinos still roamed Bubiana, but many were in danger of being killed.

This demonstrates that the continued success of the conservancies and similar projects, such as Zambia's South Luangwa Area Management Units, cannot rest on profits for the large landowners alone. To survive over time, especially in nations where despotic governments and autocratic tribal chieftains can destroy progress by fiat, it's necessary to involve rural Africans in building a system that's sustainable for them.

Communal and resettlement lands surround about two-thirds

of the Bubiana Conservancy and more than four-fifths of the Save Conservancy. These lands are densely populated and poor, yet they have been drawn into the management of the conservancies because they recognize that success will mean a better life for them and their families. Organizations such as the Save Valley Conservancy Trust function as philanthropic outreach arms, supporting local economic development efforts.

"We've learned we can't save the rhino only through our guns," said Kenneth Manyangadze, the head guide at Save, in an article for *International Wildlife*. "The answer is also through the hearts of the people."

A local council member put it this way to De Alessi: "To us, the rhino are worth a lot more alive than dead."

Another villager told essentially the same story in Zambia's South Luangwa district, where wildlife frequently roam out of the park and into private or communal lands. "Although the elephants destroy our crops, at the end of the day, we also get a better benefit from them," he said during the filming of *Conservation Pathfinders*. "By the tourists coming in, looking at them, and going back, they leave us some few monies so that we are able to do more."

Involving rural Africans in private conservancies and other private management schemes has a democratizing effect, which sometimes doesn't sit all that well with governments or tribal leaders who would prefer to keep people under their thumbs. Where bottom-up democracy can be established and tenure over resources secured, the people will have a stake in their land and will rediscover a conservation opportunity that had been taken from them once before.

"There are two principles: The first thing is to make money, to convert wildlife into dollars and cents, using tourism or safari hunting or other forms," said Brian Child, now a faculty member of the Center for African Studies at the University of Florida. "Once you've got that, the next trick is to use the money effectively, so that the communities realize the value."

The linkages must be direct and visible. A portion of the revenues from hunting permits and taxes on tourist lodges, for example, must be returned in direct payments to families and used to fund improvements in infrastructure. In the villages around the South Luangwa Park, those decisions are made democratically, with citizens often choosing to invest in game scouts, water wells, electricity, health clinics and schools, where a new culture of conservation is being taught to the next generation.

"They're putting 20 percent of their cash into employing game guards. That means they must believe a little in what they're doing," Child said. "What we try to do is turn people into responsible citizens. The shift from [tribal] subject to citizen is not going to happen overnight, but I can see evidence every day that the shift is occurring.

"Due to the changes that are taking place now, we are able to see communal benefits and individual benefits," Child continued. "People are starting to reverse from being potential poachers to conservationists. If people are conserving elephants, buffalo and lions because they're valuable, along with that come the birds and the bees and the trees, so we're actually conserving whole regions of biodiversity. In many ways, we're seeing a wildlife revolution. I think you're going to see a whole lot more wildlife in southern Africa in twenty years' time."

For landholders such as Clive Stockil at Save Valley Conservancy, the change made sense not only from a business point of view, but also from a conservation perspective.

Rancher Ken Drummond at Bubiana Conservancy put it this way: "I really enjoy giving nature a chance. I think, at heart, farmers are fundamentally conservationists—especially if we're able to demonstrate to them that conservation can pay. That's been one of the big weaknesses in conservation for too long. Conservation has been propped up by donations out of people's handbags. That's just not a deal. You can't build an industry like that."

You can build a conservation industry, however, by giving landowners and other citizens incentives to do the right thing for themselves as well as the land. But the African experience is by no means secure. Political instability, misguided "land reform" schemes, corruption and opposition from mainline environmental groups that oppose fee hunting could combine to roll back progress. For now, however, the black rhino, elephants and other threatened African wildlife are enjoying more safe harbors. And those wildlife harbors are safer than ever when the benefits are tangible and local.

Here is how David Hulme and Marshall Murphree summarized the progress of community conservation in Africa in *African Wildlife and Livelihoods:*

1. Community-based conservation must be locally driven rather than state-centric. "No longer should rural Africans be seen as degraders of the environment but as local heroes."
2. Sustainable development means advancing the notion that the things to be conserved—wildlife and their habitats—are resources to be properly exploited and managed. "Wildlife utilization, rather than wildlife preservation, might be best for conservation."
3. Markets must play a greater role in shaping conservation incentive structures. "Following the dictum 'use it or lose it,' these notions suggest that if species or habitats are to be conserved, then they must not be isolated from the market. Rather, they must be exposed to it as their uniqueness and scarcity lead to high valorization and thus promote conservation."

Far to the north of Zimbabwe in Tanzania, close to the equator, collaboration between villagers and safari companies demonstrates the potential for community-based conservation, sustainable development, and use of markets. Together they are securing wildlife's future and building a better life for those people who don't just enjoy the big game of Africa, but must live with it.

Sinya, a Maasai pastoralist community in far northern Tanzania, holds a large expanse of wildlife habitat between major parks, including Amboseli National Park across the border in Kenya. It is here that the new conservation ethic is taking hold. In a 2004 publication from the International Institute for Environment and Development, Fred Nelson wrote that in Sinya and other villages of the region, wildlife began a resurgence when poaching was brought under control and community-based ecotourism started to deliver meaningful benefits to the villagers. There are still institutional confusion and policy perversions that need to be overcome, but the future for wildlife and people together is quite hopeful. Sinya village's tourism rose substantially and quickly in just the first few years of the twenty-first century. As Nelson wrote, "Individual income opportunities from tourism operations—which have evolved into a permanent luxury camp—have also developed through employment and purchase of local crafts." The latter aspect of ecotourism is a critical means to put cash into the hands of Sinya village women.

Nelson continued, "As in Ololoskwan, tourism income has created important village-level incentives for wildlife conservation in Sinya. In just a few years, the area's large mammals have gone from being a costly resource that the community had co-existed with over the years to a source of valuable collective income and individual employment. The income has been used for conventional social service infrastructure priorities in Sinya, most notably construction of the primary school dormitory and maintenance of water supply machinery."

Elephants are on the rebound in Sinya, and their increase—within reason and before habitat destruction, which elephants can wreak if overabundant—can help villagers earn more money from ecotourism. At this time there is no legal hunting of elephants in that area, but elsewhere in Africa, elephants can be hunted legally and revenues invested locally. Earning money from elephant hunting is an effective conservation strategy that conserves not only elephants but also the thousands of other animal and plant species that use the same habitat.

Early in 2005, ABC television's *20/20* news program illuminated a fact that might trouble animal-rights activists but is a comfort to those who wish to see a future with wild elephants. African nations that permit elephant hunting are enabling elephant numbers to rise; nations that have banned hunting may have imposed a feel-good law but have all shown declines in their elephant population. The reason is simple: Hunters who pay thousands of dollars to kill an elephant in a controlled hunt leave those who work on the hunt and others who have an ownership stake in that animal much better off. As is true of so many other resources, "Wildlife that pays, stays—and grows."

A summary monograph by Fred Nelson provided specific evidence supporting his thesis that "Safari hunting is essentially a small but extremely lucrative component of Africa's burgeoning tourism industry."

- The trophy fee for a single elephant in southern African countries is about $10,000 to $15,000 in U.S. currency.
- In Zimbabwe, private landholders were granted broad user rights to wildlife. Since 1975, 27,000 square kilometers of that nation's private lands have been placed under management for wildlife. Animal populations rose fourfold in less than a single human generation.
- Within Namibia, a single lion or elephant hunted at the Torra Conservancy yielded $10,000 just for the trophy fee. The community got that money directly.

The counterexamples continue to dominate, however. Local disincentives stemming from Tanzania's central-government approach to safari hunting induce a decline in wildlife and loss of vital revenues to thousands of Tanzanian citizens in the rural areas.

Community conservation in Africa isn't as much about changing laws or techniques as it is about changing attitudes and social order. Only as African nations move further away from the command-and-control legacies of the colonial and, more recently, the totalitarian eras will community-based conservation gain broader acceptance. Secure tenure in resources and bottom-up democracy are essential elements of a civil society that benefits people and wildlife.

Community-based conservation has dedicated adherents in parts of Africa, but those disciples of change will need time to spread their new and exciting gospel. It took generations to degrade Africa's wildlife richness; it may take as many to rebuild it. But the evidence is overwhelming that rebuilding will be done when the people on the land with the animals can secure benefits from it.

15
Letting a River Be a River
Learning from a partnership in undamming a river's natural potential

There are 2.5 million dams of all sizes on American rivers and streams, which means it's inevitable that many communities will face a choice between keeping and maintaining their dams—or removing them and restoring the river or stream to its free-flowing state. Here's how groups in Wisconsin got started on the nation's largest restoration project.

"MEN MAY DAM IT AND SAY THAT THEY HAVE made a lake," conservationist Wendell Berry once wrote. "But it will still be a river. It will bide its time, like a caged animal alert for the slightest opening. In time, it will have its way; the dam, like the ancient cliffs, will be carried away piecemeal in the currents."

Americans have a love-hate relationship with dams. They're a part of our history and our culture, reminders of a time when rivers were the superhighways of a new land. Well into the twentieth century, dams were ribbons of commerce and transportation, powering grist mills and, later, small electric plants. Today, many dams have outlived their usefulness. They can pose safety hazards, damage fisheries, economically burden their private or public owners, and generally prevent rivers from executing their natural functions. In many communities, however, nostalgic ties to dams are strong. They can be picturesque symbols of community identity and a source of civic pride long after their useful lives have expired.

In the rolling hills of central Wisconsin, private citizens, conservation foundations, local governments and the state Department of Natural Resources have bridged the gap between the romance of what dams once were and the reality of what's best for their communities and the river today. They did so through a six-year project that may have seemed to culminate in October 2001 along Wisconsin's Baraboo River, where the last of four dams that had impeded the flow of the river since Wisconsin's territorial days was removed.

But the tangible benefits to the people who live along the Baraboo and have responsibilities through their municipalities, businesses and recreational interests are flowing out of unglaciated hill terrain and springs, into small feeder creeks and then through the now free-flowing mainstem of the Baraboo River itself. Put simply, those benefits are lowered safety risks, reduction in costs to city government and thus a lower tax burden, increased opportunity for responsible economic development near the river, improved water quality, more recreational use of the river and, most astoundingly, a resurgent community of native fish including lake sturgeon and smallmouth bass.

Today, a leaf that wafts into the water at the Baraboo's headwaters near Elroy may meander all the way down the 160-mile river to where it meets the Wisconsin River near Portage. That makes it the longest mainstem of a river to be undammed in the United States—and an environmental precedent that will be closely studied for years to come.

The last dam to be removed was the Glenville (also known as Linen Mill) Dam just downstream from Baraboo, which is about thirty-five miles northwest of the state capital of Madison. Although some people along the river had opposed breaching the dams, others recognized that a free-flowing Baraboo River would be better for people, aquatic life and, ultimately, the economy of river communities.

The long and necessary process of convincing people that the dams should be removed fell largely to the River Alliance of Wisconsin and the Sand County Foundation, two groups that practice a hands-on style of environmentalism by finding solutions that work for nature as well as landowners. Thanks to the efforts of these groups and other private organizations working with landowners and governments, attitudes about the dams changed as people learned more about their pros and (mostly) cons. In addition to being dangerous to boaters and others who use the river for recreation—two people drowned at the Glenville Dam in July 2001, just a few months before the hazard was eliminated—the dams prevented fish migration and increased siltation.

It will take scientific monitoring to measure the effects over time, but apparently the free-flowing Baraboo is quickly reverting to the river it was before man built dams to harness the power of water to run his mills. A fish survey made before the removal of the LaValle Dam in February 2001 showed thirteen species of fish above LaValle, with carp being the most common. Six months later, twenty-six species of fish had already congregated at the former dam site, with smallmouth bass being the most common species. More species are once again using river stretches from which they had been isolated for decades. The dam's rocky riffle and plunge pool are being maintained by the state Department of Natural Resources so that fish needing this type of habitat can continue to spawn there.

A free-flowing river means more oxygen and less nitrogen in the water, which is another reason why fish will be the primary beneficiaries over time. Age-old breeding patterns may be re-established as fish swim upstream to spawn. Each spring, the river will do what comes naturally: It will rise quickly, recede quickly and leave nutrients in the flood plain, thus enhancing those biotic communities that thrive in riparian areas.

By the way, fewer nutrients being dumped into the Wisconsin River means less trouble for the Mississippi River and the so-called "dead zone" at the mouth of the Mississippi in the Gulf of Mexico. Early in 2004, the Baraboo River was removed from the list of Wisconsin's "impaired" waters. The monitoring and research associated with the dam removal partnership shows on an objective basis that recovery of a natural asset can be good environmental business and good for businesses at the same time.

In the city of Baraboo, the undamming of the river may unleash the local economy as well. Plans are percolating to revitalize a part of the city's downtown that fronts on the river. The nationally known Circus World Museum is already within sight of where one dam was removed, and a developer envisions turning back the clock to recreate a new gateway for one of Wisconsin's oldest cities. The Nature Conservancy and the Aldo Leopold

Foundation already have offices in Baraboo; now there's talk of a Leopold Museum. The city may also try to attract a mix of year-round shops and businesses tied to the environment, health and outdoor recreation.

The Baraboo is only one river with all of its currently unsafe, uneconomic dams to be removed, but it may serve as an example to community-based environmental groups all across the nation. By purchasing obsolete and unproductive dams and working with landowners as well as local, state and federal agencies, groups such as the River Alliance of Wisconsin and the Sand County Foundation have shown it's possible to accomplish environmental good without name-calling, lawsuits or excess regulation.

Others can follow this model. The River Alliance of Wisconsin and Trout Unlimited have published a step-by-step guide that captures the principles of assessment, planning, execution and evaluation. *Dam Removal: A Citizen's Guide to Restoring Rivers* is an eight-chapter toolkit. It begins with the premise that thousands of communities, small dam owners and government agencies are facing—or soon will face—the question of what to do about a dam.

Chapter One explains what dams are and what they do. It introduces the reasons why people may be concerned about the effects of dams on rivers, and explains why many dams must be repaired or removed.

Chapter Two provides general information about the dam repair or removal decision process, focusing on what role citizens can play. It introduces the four stages of dam removal: planning, draw-down or draining of the impoundment, removal of the structure and postremoval environmental restoration.

Chapter Three continues to hammer home the importance of assessment. Before a rational decision can be made, it's vital to learn as much as possible about the dam itself. What kind of dam is it? Who regulates it? Who owns it? This chapter is an organizational map, complete with a worksheet for recording information.

Chapter Four is a primer on how to conduct research about dams, also key to assessment.

Chapter Five tells citizens what to do with the information they've gathered. Most of that data will fall into four categories: environmental, economic, engineering and societal. Also, four themes emerge around dam removal projects as those issues inter-weave: fish and wildlife, land and property, public safety, and sediment accumulation. Knowing how these issues connect is cru-cial to the planning process.

Chapter Six lists the tools that can be brought to the plan exe-cution or action stage. Those tools are largely economic, legal and environmental.

Chapter Seven teaches advocates of river restoration how to win the public-relations and decision-making campaigns that may well involve a mix of public bodies and private property owners. What are the best strategies? What are the right tools? Who are the right stakeholders?

Chapter Eight is about what happens once a decision is made to remove a dam. Here, the guide makes the important point that the work isn't done with the removal of the dam. "Don't forget your goal is river restoration, not dam removal.... It's time to restore the river and reconnect the community to a free-flowing and healthier river." In other words, execution and evaluation are sometimes joined at the hip.

The work of the River Alliance continues on related fronts. In partnership with the Wisconsin Academy of Sciences, Arts and Let-ters and the University of Wisconsin at Stevens Point, the River Alliance has coordinated a "Waters of Wisconsin" initiative to focus on non-point strategies. The alliance is also part of the Groundwater Quantity Working Group, which includes some of the same partners plus the state's vegetable growers. The goal? "To recommend strategies to protect the state's groundwater resource while maintaining the economic sectors that depend on it."

Some 2,500 years ago, the Greek philosopher Heraclitus cap-tured the ever-changing essence of rivers when he observed, "You can't step in the same river twice." Likewise, no two dam removal and river restoration projects are likely to be the same. However, lessons can be learned from those who have already gotten their hands wet in freeing a river.

To obtain a copy of *Dam Removal,* contact the River Alliance of Wisconsin, 306 E. Wilson Street, Suite 2W, Madison, WI 53703. Telephone: (608) 257-2424. Internet: www.wisconsinrivers.org.

16
Going Down to the Sea in Ships
How fishermen might be able to develop greater responsibility for fisheries and secure safer working conditions

Those who have chosen the profession of fishing venture out in some of the most dangerous conditions to produce some of the most nutritious foods available—seafood. The challenge in sustaining fisheries is not just to manage the hazards of the profession but also to demonstrate responsibility within a hostile policy environment. But across North America, fishermen in small groups and in coastal communities are showing the way to healthier fisheries and more vibrant human communities.

IF IT IS BETTER TO TEACH A MAN TO FISH than to provide him with fish, it should also be better to let a skilled fisherman make his own choices, informed by experience. But instead, we seem determined to let bureaucrats hundreds or thousands of miles away from the fishing grounds make choices for the fishermen who must live—or die—with the consequences.

It is becoming evident to those who study the issue that centralized, top-down regulations are making the human risks associated with fishing—already a dangerous profession—much worse. In the bargain, the command-and-control approach is also making conservation of fisheries much harder. A newspaper article published in Arizona on January 13, 2005, and a fishermen's conference in California that began on the same day combine in an unlikely way to help tell the story.

Writing in the *Tucson Weekly*, Renee Downing noted that "Fishing is a dangerous trade anywhere; in the North Atlantic in winter, the risks are horrendous.... The sinking of the Northern Edge off Nantucket (just before Christmas 2004) was the worst loss of life in the New England fishery since the Andrea Gail of Gloucester, Mass. went down with all six hands in 1991."

The article continued, "There was a lot of anger in New Bedford. Federal fishing regulations had kept some scallop beds closed all summer.... Officials only opened them in November, when the weather was turning treacherous.... One set of rules imposed penalties on boats that came into port for any reason before their time in the restricted waters was up."

CBS News reported on the incident on December 22, 2004, within a short time of the capsizing of the *Northern Edge* as captain and crew had been scalloping for the winter holiday trade. Part of the transcript of the news broadcast reported, "Fishermen are allowed a single trip into the area through January, and can catch a maximum of 18,000 pounds over 12 days."

A leading New Bedford fisherman, Rodney Avila, a member of the New England Fishery Management Council, declared that the rules force fishermen to stay at sea in bad weather to take advantage of the limited opportunity, according to the *CBS News* piece.

On the same day that the Downing article pointed to the same conclusion, several dozen fishing representatives spanning the coasts of North America met, most for the first time, at a workshop in Del Mar, just north of San Diego, sponsored by the Sand County Foundation. A wide array of fish resources and the disparate time zones of their home ports did not dissuade these people from reaching some general conclusions. According to an article by Becky W. Evans in the *New Bedford Standard-Times*, "Conference participants criticized federal management policies that have shortened fishing seasons, sent fishing boats out to sea in dangerous conditions and created fishing derbies that lower fish prices by flooding the market with the same product."

But it doesn't have to be this bad. It can even get better. One need only look to New Zealand, where reforms of natural resource policy and reaffirmation of treaties between the native Maori people and the United Kingdom have led to boats chasing and then catching more fish. And more people are profiting from marine resources in which they have a literal ownership stake. The device now widely used in New Zealand is that of individual fishing quotas—a transferable right to a designated fish stock. Another approach that restricts entry and thus improves chances for conservation is permanently assigning to a fisherman or a group of fishermen a designated area of ocean. Fishermen in that Southern

Hemisphere country can keep their boats at the dock in the worst weather. And they can wait for market demand to increase prices before they cast their lines free of the pier.

In contrast, an Alabama red snapper fisherman named David Walker reported that, "Under current regulations, [Gulf of Mexico] fishermen can only catch depleted red-snapper stocks in frantic 10-day spurts that create dangerous conditions for both fishermen and fish."

While fishermen don't necessarily agree that designated property rights are an appropriate resource management approach, they did agree, at least at the Del Mar gathering, that it is possible to produce policy reforms and changes in practices that will benefit both fishermen and fish. Transforming the government's role from regulator to facilitator and assigning greater responsibility for stock management, and even recovery, to the fishermen would be desirable.

Within the United States, a burdensome, expensive and convoluted mass of policies, rules and regulations confine the opportunity for fishermen to cooperate even for the critical purpose of restraining take so that fisheries can grow back to levels that could sustain profitable harvest.

Antitrust legislation has been interpreted by federal courts to rule that many cooperatives of fishermen are illegal. The Magnuson Act stifles the creativity that fishermen could bring to resource stewardship. Very large agencies with thousands of employees in the federal and many state governments impose penalties and produce very few incentives.

There are bright spots, however, even in the United States, where a number of national environmental groups continue to push for even greater hegemony by the federal government over fisheries management. There is a possibility for reform of the Magnuson Act. This could bring greater experimentation and real-world testing through adaptive management of fishermen-led conservation.

Another hopeful spot is the leadership of the State of Alaska in fisheries reform. Some of the greatest cold-water native fisheries in the world exist in Alaskan waters. These would be sustainable sources of protein and profit well into the foreseeable future if watersheds kept their intrinsic productive qualities that support spawning fish like the famous Copper River salmon. In the face of markets glutted with pen-raised Atlantic salmon, Alaska fishermen have fallen onto the double swords of a collapse in prices and a rise in operating costs as fuel expenses have risen sharply. To the credit of many politicians and bureaucrats in the forty-ninth state, they have established a more favorable environment for utilization of wild and sustainable fisheries. Some innovations, such as the Chignik sockeye salmon cooperative, have been allowed and show some tentative promising results.

There is no one-size-fits-all approach that will sustain fisheries and retain healthy human communities. Conditions are too varied; too many fishery stocks will need decades to recover before they can be profitably harvested; and the sea itself is subject to massive changes. But a few things are true and universal when it comes to fishery conservation. One is the saying, "What can't last won't last." Permanent losses due to overfishing are evident around the world through the tragedy of the commons. What is owned by all is a thing that can't last. Another truth is that among fishermen, hope abides. There will always be people who go down to the sea in ships to try to catch the fish dwelling there.

A scientist who attended the Del Mar conference and has a wide range of research experience on fishing in the Atlantic Ocean seems to understand these universal conditions. Dr. Kathleen Castro of the Rhode Island Sea Grant program put it this way in Becky W. Evans' article in the *New Bedford Standard-Times:* "There has been a belief that the fox can't guard the hen house. But the fox can guard it, grow it, harvest it, and sustain it. It needs to be stated that fishermen can do it. There are successful cases of fishermen self-management."

One can only hope that policy reforms within U.S., Canadian and Mexican agencies will lead in the direction of greater fishermen self-management and realize more opportunities for the fulfillment of Dr. Castro's assertion. For the benefit of the fishermen, their families, and the fisheries upon which their livelihoods depend, an increasing amount of self-governance would be a very good thing.

17
Lessons Learned

Across the United States, in Africa and beyond, citizens who care to be responsi-
ble, who can benefit from resource use and who are able to participate in
bottom-up democracy are choosing to become indigenous stewards of the land.
They are neither relying upon nor waiting for government- or court-imposed
solutions. For many others, however, there is a dark despair. Among some
activists there is even a feeling that environmentalism has died. What has pro-
duced such different perspectives?

IN THE LATTER MONTHS OF 2004, an essay by Michael Shellenberger
and Ted Nordhaus titled "The Death of Environmentalism"
crossed many people's desks. Around the world it showed up on
the computer screens of environmental activists. While the article
focused on one issue, global warming, the authors believe there are
lessons to be learned about the failure to impose a policy-mandated
type of solution to this challenging problem. They are not alone.

Onetime Sierra Club president Adam Werbach declared that
environmentalism is in distress, has lost much of its power, and as
it is currently manifest may be incapable of producing solutions to
the most serious environmental troubles. Mr. Werbach has gone
so far as to promise never again to apply the environmentalist label
to himself.

Despite the worries, environmentalism is not dead. But to live
anew, it must take a different direction, embracing an ethic that
expects the best of men and the land.

Much of environmentalism has been, simply put, hijacked
through inappropriate use of command-and-control, undue
reliance on legislative solutions, and excessive zeal in regulation
by government. Gregg Easterbrook, a noted author with strong

environmentalist credentials, has established that many of the well-intentioned water quality programs of the Environmental Protection Agency are not performing well but will be highly resistant to reform. In his opinion, threats of lawsuits and intensive lobbying to support and to alter legislative devices like the Clean Water Act will prevent any meaningful improvement. Reform of the Endangered Species Act to incorporate incentives for landowners rather than punishment may be desirable; but the ideological battle lines are sharply drawn, and the battlements on each side are heavily defended.

If only some of that money for lobbying congressmen and senators could go instead to incentives for landowners to improve or restore wildlife habitat.

In the United States and Europe, much of the command-and-control form of environmentalism is an outgrowth of New Deal legislation, socialist inclinations and large government agencies developing a life of their own with an inexorable desire for hegemony. Those are all a long way from the practices of people who in the mid twentieth century took Aldo Leopold's example of hands-on land restoration seriously.

Of course the situation is worse for people striving to live a more environmentally responsible life in dictatorial and totalitarian nations. Those many Kenyans who embraced community mobilization efforts to plant trees, establish gardens and secure more meaningful uses of resources, and did so at cross purposes with Daniel Arap Moi, felt the sting of his lash in both a literal and a figurative sense. Fortunately, citizen leadership is improving both democracy and the environment even in Kenya.

There's no correspondence course to teach people an environmental ethic based on incentives and on community and individual responsibility. There are no "bachelors of community-based conservation" in the nation's leading universities yet. (Perhaps, however, the Swedes are leading the way with a recently established interdisciplinary master's degree program in ecosystem management at Stockholm University, which emphasizes an understanding of the complexity of the social-ecological systems within which we live.)

There are no "Sand County" blueprints hanging on the walls of America's leading environmental organizations, and no hands-on environmental planning books on the shelves of the nation's top corporations. In part, that's because people have only recently begun to find their way back to Aldo Leopold's timeless lessons. But there's another, more basic, reason why a playbook doesn't exist—and may never be written—for the movement that is quietly displacing the state-sponsored environmentalism of the late twentieth century. Simply put, in a world where the antidote for misapplied top-down regulation is bottom-up partnerships in which many people can help shape the solution, there are fewer hard-and-fast rules. What works in Pennsylvania may not be effective in Oregon—and may be entirely different still from what works in Illinois. As the pioneers of community-based conservation are learning, replacing the "one-size-fits-all" environmental philosophy of the past means tailoring solutions to fit the sizes, shapes and styles of twenty-first-century communities.

This also means that there will continue to be human health considerations, ecosystem viability matters and even biological diversity concerns that are going to require national effort. There will continue to be situations where large, reasonable and centralized environmental solutions are necessary.

Among the great imperatives is for national governments, and state or provincial governments in federalized nations, to build upon the legitimate desire of their citizens to improve environmental quality and natural resource integrity and value in their communities. This means "cutting the cloth" quite differently from the one-size-fits-all patterns of devices like the U.S. Clean Air and Clean Water acts.

There will be benefits gradually and affordably with such a change in emphasis. One that will be welcome to the governments that embrace bottom-up solutions to a greater degree is to be able to cut taxes by doing so. When environmental leaders such as Paul Portney, president of Resources for the Future, can point out that regulation of water quality in the United States now costs tens of billions of dollars annually and is largely a maintenance expense for preserving the costly status quo, it is likely that money which should be helping to improve water quality is not being used effectively.

Tax revenues that are used poorly are just one cost that can be trimmed. The government-mandated approach induces severe social and human costs. These, too, can be trimmed. Throughout this book we have made it clear that reducing barriers to those who wish to develop greater responsibility for rare species, for water quality, for a more viable fish community, for forests with less likelihood of catastrophic fire and the like will reduce social distress. Building in responsibilities at the local level may even keep more fishermen alive longer.

On the one hand, landowners and citizens in communities can be challenged and will listen when they are told about the need to do something to help wildlife, let's say. They will respond and be fully engaged when they can lead the way in developing meaningful, affordable solutions while also capturing some of the benefits of improved management. But instead, many governments have used the clout of law to inhibit the development of local responsibility and to stifle local innovations.

One species of great newspaper headline value, the attractive and thrilling gray wolf, exemplifies the distinction between approaches that engage citizens and those that enrage citizens. While there are clear justifications at the ecosystem level as well as advantages for biotic diversity in restoring gray wolves to the Yellowstone Park area of Wyoming, Idaho and Montana, there also are resentment and conflict and severe social tension around this issue. The U.S. Department of the Interior mandated wolf release in the early 1990s, so it was done and the fires of conflict were stoked.

In contrast, a partnered arrangement between the Department of the Interior and the State of Wisconsin was built in the early 1980s. It relied upon several features for its success and headline-avoiding good work: substantial engagement with potentially affected groups and solicitation of their help in allowing wolves to return; providing a strong role for private groups in fundraising and education; payment for wolf depredation—much of this going to reimburse owners of wolf-killed dogs—and giving assurance that with good habitat and an overabundant prey source, white-tailed deer, wolves would take matters into their paws. A wolf pack now lives within fifty miles of the Wisconsin state capital. Hundreds of wolves cause few and mostly manageable problems. The stage is set for continuing improvement in gray wolf management if the responsibility for wolves can be set, appropriately, at the state level. If nationally mandated management of gray wolves were really appropriate, wouldn't we expect a wolf pack to be residing someplace in Maryland or Virginia, close to Washington, D.C.?

So it is that *New York Times* articles in 2004 acclaimed the largely positive Wisconsin model of wolf recovery, while resentments, feuds and threats of legal battles still waste money and make people very unhappy in the northern Rocky Mountains.

Lesson One: Make sure that your tent is big enough to shelter everyone who needs, or wants, to get out of the rain.

COMMUNITY-BASED CONSERVATION EFFORTS don't work unless they can accommodate the range of players in the community. When the Trout Creek Mountain Working Group first came together in southeastern Oregon's arid, high-altitude cattle country, it was an unlikely, often cantankerous, collection of people. There were ranchers who didn't want to change 130-year-old grazing practices. There were environmentalists who saw no reason to compromise on their demands to stabilize denuded riparian areas and to protect the rare Lahontan cutthroat trout. And there were regulators who saw no advantage in listening to anyone other than their own bosses. The only thing they had in common was a fear of losing in a protracted confrontation. So they decided to help everyone win.

Present for the first meeting of the Trout Creek Mountain Working Group were representatives of the Whitehorse Ranch, the Izaak Walton League, Oregon Trout, the Oregon Cattlemen's Association, district managers for the federal Bureau of Land Management and a mix of state natural resources officials. "The tension, energy, fear, care and concern in that room for four hours was overwhelming," recalled "Doc" and Connie Hatfield, owners of Hatfield's High Country Ranch and two of the group's founders. "At the end of the day it was obvious that changes had to be made or everyone, and the land, was going to lose big after a long battle in court. Regardless of the grazing decision made by [the Bureau of Land Management], ranchers or environmentalists were going to challenge it with a lawsuit," which would have guaranteed several more years of stalemate.

In short order, other ranchers and the Oregon Environmental Council joined the working group. Because the right people were at the table, and the regulators were willing to give a community-based, scientific solution a chance, the group set out to make an immediate difference. Voluntary changes in grazing practices were instituted for three years while sustainable land and resource management strategies were developed. The result has been a grazing rotation system that works for the ranchers—and cleaner, colder streams for the trout.

"Major, positive, documented changes on the land are a reality for everyone to see today," said the Hatfields, who wrote a history of the Trout Creek Mountain Working Group. The couple run about four hundred cattle on a 25,000-acre ranch that includes private as well as public lands. "It takes people to improve land. We already have more laws and technical information than we need.... The time is right for more 'people-to-people' alliances where landowners, environmentalists, and state and federal agency folks work cooperatively to produce action on the ground."

Murray Lloyd of Shreveport, Louisiana, a founder of the Black Bear Conservation Committee, couldn't agree more strongly. "The [command-and-control] system is set up for conflict. We have been successful in creating mechanisms...to allow discussion in non-combative ways," he said. "By putting faces and personalities to 'those' people, we have avoided a great deal of name-calling and rhetoric, allowing healthy debates with differences of opinion but without personal attacks."

After attending a seminar hosted by the Sand County Foundation, Lloyd began work on what came to be called "The Wild Goose Chase," which later evolved into a virtual regional group called the Conservation Network. The network has brought together industry, government and conservation groups to "create workable solutions for species and habitat conservation" in the lower Mississippi Valley. It provides yet another example of how partnerships can work.

Lesson Two: Remove the barriers that separate dirty little problems from green solutions.

BUREAUCRATIC UNRESPONSIVENESS, economic disincentives, miscommunication, lack of good data and deliberately obstructive processes often stand as obstacles in the path of community-based conservation. Every hands-on environmentalist encounters them; the successful ones overcome them.

Consider the wildlife protection efforts of Environmental Defense, which has restructured its endangered species programs by substituting market incentives and protection of landowners against the Endangered Species Act's perverse incentives for command-and-control thinking.

"For the past seven years, we have devoted most of our time and attention to the conservation of endangered species on privately owned land," said David Wilcove, senior ecologist for Environmental Defense (www.environmentaldefense.org). "In particular, we have focused on the use of economic and regulatory incentives to encourage private landowners to restore habitats for endangered species. By using 'safe harbor' agreements, mitigation banks and financial incentives, we have been able to get landowners across the country to proactively manage their properties on behalf of endangered species—actions that would have been unthinkable a few years ago."

About 1.5 million acres of privately owned land have been enrolled in "safe harbor" programs, which typically provide technical as well as financial help for landowners who are willing to make their land attractive to endangered or threatened species. The Landowner Conservation Assistance Program of Environmental Defense has enrolled 61,000 acres in the Texas Hill Country alone, Wilcove said, restoring habitats for two endangered birds and performing "beyond our expectations."

What barriers typically stand in the way of success? Usually it's not the landowners themselves, according to Wilcove, but regulators who don't see the wisdom of trying something new. "We found that one of the key determinants as to whether a 'safe harbor' program will succeed or fail is the local staff of the U.S. Fish and Wildlife Service," he said. "If someone at the local office recognizes the value of this approach and is determined to make it work, then, in all likelihood, the program will succeed. But if the staff is skeptical about the value of 'safe harbor' programs, then the program will, in all likelihood, not get off the ground."

Also, good intentions are no substitute for tangible help. "A second important obstacle is simply staffing and funding to make these programs work. Landowners want information on habitat improvement; they want to participate in these programs," Wilcove said. "What is often lacking are sufficient resources to take advantage of their interest."

Harold "Bud" Jordahl, a cofounder and president of Wisconsin's Gathering Waters Conservancy, said a common obstacle in dealing with landowners is that they sometimes back off for fear of compromising their own property rights. It's often hard for landowners to switch from years of being on the defensive against regulation to embracing partnerships that require trusting a public agency. "Property owners feel strongly about their rights and have, all too often, little or no concern regarding their responsibilities as owners," Jordahl said. "However, we all have responsibilities for conservation...and private responsibility is a refreshing contrast to the use of only one tool, regulation."

California rancher Pete Stent said the "permit maze" and other complex regulations often stand in the way of private landowners or community-based conservation groups devising their own solutions. Consequently, landowners are concerned about their legal liability if they take a chance on a process or project that doesn't work as planned.

"Paper, forms and [a preoccupation on process] are the mainline defenses" for the command-and-control environmental movement, agreed Emmer Shields Jr., the highway commissioner for northern Wisconsin's forest-covered Ashland County. "Process becomes a means of obstruction. Another aspect of regulation is complexity, which leads to avoidance or outright refusal to comply. Allies will not be created by pointing a regulatory gun at them."

When regulations are easy to understand and consistent with a community's land ethic, they will be observed. Hands-on environmentalism cannot take root when the very people who want to bring about change aren't able to get their hands around the problem.

Lesson Three: Understand the difference between conservation and preservation.

Three decades of efforts to save Africa's wildlife taught Graham

Child, one of that continent's longtime conservationists, the important difference between "protecting" threatened animals and
ecosystems through punitive measures and "conserving" them by
giving people an incentive to safeguard the land. "The main barrier
we had to overcome was the dogma entrenched in conventional
conservation—actually, protectionism," he said.

Wildlife protectionism in Africa has led to problems that range
from poaching to habitat destruction to government neglect of natural areas. The preserved colonial practices of denying people the
use of their own land for fear it would be plundered of wildlife has,
ironically, led precisely to that result. Child's work and undertakings like the thirty-year-old CAMPFIRE program have tried to
re-establish the link between people, land and wildlife.

A Zimbabwe initiative launched in the early 1970s has led to
similar programs in a half-dozen other African nations. The principle is simple: Given some carefully balanced rights to use wildlife
under free-market conditions, people will conserve not only the
wildlife but the habitats that protect it so they can sustain their
livelihoods.

The problem, Child explained, is that African governments and
tribal leaders don't always appreciate the democratizing effect of
giving people and landowners more control of resources. "Totalitarian governments see this as a threat to their power, rather than a
factor promoting their popular influence," he said.

Half a world away, in the pine forests of northern Wisconsin,
Ashland County's Emmer Shields also stresses the difference
between preserving resources and conserving them. "Conservationists are hands-on people," he said. "For the most part,
preservationists subscribe to a more regulatory approach to environmental protection." Many times, preservation is much more
costly than conservation—and it stands in the way of true progress,
Shields added. "I can't tell you the number of times that I've seen
huge expenditures of effort go into saving trees that are near the
end of their lives. If that same effort was put into planting trees to
replace the ones being lost, the world would be a better place. Too
often I see opportunities for enhancing or bettering the environment trashed in order to preserve a questionable resource, such as
a low-functioning wetland..... Rigidity of regulation and failure to

empower regulators quite often leads to high costs and little benefit to society."

Lesson Four: Be patient until you can't be patient any longer, then be patient some more.

IT TOOK YEARS TO CREATE SOME of the environmental challenges confronting the world, so meeting those challenges might take more than a few weeks, too. "First, recognize that it takes time—years. Second, develop a long-term strategic plan. Third, stick with it," said Bud Jordahl, whose Gathering Waters Conservancy in Wisconsin is a coalition of land trusts formed in 1995 to assist trusts, landowners and communities in their efforts to protect land and water resources (www.gatheringwaters.org).

In more than fifty years of conservation work that has included working for Wisconsin's Department of Natural Resources, for the U.S. Department of the Interior and as a professor of urban and regional planning at the University of Wisconsin, Madison, Jordahl has never lost faith in the ability of citizens to do the right thing if given information and time. "Recognize there is a latent but potentially strong base of support in the average, common citizen. Exploit that!" he advised.

Others counseled taking the time to hear what people have to say. "First, it is important to spend time listening to the landowners to understand their desires, concerns and fears," said David Wilcove of Environmental Defense. "Unless these are addressed, the program is unlikely to succeed."

It took five years for the Trout Creek Mountain Working Group to see any tangible results in Oregon's cattle country. From its tense beginnings, the group evolved into an organization with rules of engagement that made meetings constructive. But having the patience to stick with an inclusive process was crucial, as Doc and Connie Hatfield recalled. "After everyone's voice enters the room, two or three 'opportunities in disguise' (more commonly known as 'significant problems') are discussed. This is in the circle as a whole, or in smaller breakout groups, but always with each person given the opportunity to speak in turn and to be listened to with respect. If we don't vent, nothing happens. We have learned how to vent."

Another point that may seem self-evident outside ranch country was key to the success of the Trout Creek Mountain Working Group: ranch wives were included in the circle. "Ranch men frequently are bound by tradition to the way it has always been, which makes opportunities for change difficult to see," said Connie Hatfield. "Women in general tend to be...better able to understand the feelings of environmental folks who are viewing the situation from a much different perspective from the ranchers."

In the thirteen years since the Trout Creek Mountain Working Group was formed, change has been constant. There has been personnel turnover in the federal and state agencies involved, as well as in the environmental groups, so each new person needs to learn what everyone else already knows. The courts rejected a legal challenge filed by people outside the group, but there's no guarantee that someone else won't file a lawsuit again.

"Patience is the whole thing," Connie Hatfield said. "It's very important, and it's awfully hard. But because we've been patient, this small group of humans has helped to heal a scar on the mountain that took more than a hundred years to create. On the mountain, they're back to catching [Lahontan] trout and the ranchers are doing fine, too. We're quite honored that we can run cattle on this land, and we'll do our best to take care of it, too. In fact, we wouldn't have it any other way."

So it has come to pass that a Sierra Club president of recent years, Adam Werbach, has said he will abandon environmentalism. Meanwhile, landowners like the Hatfields in Oregon are embracing hands-on environmentalism in order to operate more profitably and improve the land and water where they live and work. Is there any more striking illustration of the aphorism "What can't last won't last" than this contrast between a person who invested himself heavily in government-mandated solutions and seems to give up in despair, and citizens who know there are challenges ahead and will keep moving on to meet them successfully.

18

Kissing a Toad

In East Texas, a tiny endangered toad is having a
princely effect on private land conservation

*Development pressure and the resulting loss of habitat had all but wiped out
the Houston toad, but thanks to a Safe Harbor agreement, a community-based
conservation ethic and the efforts of one committed rancher, the toad may
make a comeback.*

ASK BOB LONG TO DESCRIBE WHAT A Houston toad looks like and
his answer will take a few moments to compose, but ask him what
it sounds like and you will get an answer in an instant. That is
because the birdlike trill of breeding Houston toads is on its way
back to being a pleasant tune to landowners who are investing
themselves in the recovery of this endangered species.

Bob Long and the other members of Bastrop County's Houston
Toad Conservation Project Work Group have been doing their best
to ensure that future generations have a chance to hear (and see)
this small nocturnal toad in the habitat where it once flourished.
Long and his family entered their 550-acre cattle ranch in a ten-
year "Safe Harbor agreement" to help save the Houston toad from
extinction. It's a story—and a process—that is being repeated
across the United States by private landowners who are demonstrating that incentives, not regulation, are the best way to save a
species on the edge of being lost.

The Houston toad isn't much to look at, even among the larger
family of toads. It ranges in size from two to just over three inches
long, and its color may be light brown or gray or purplish gray,
sometimes with green patches. Males have a dark throat, which

appears bluish when distended. It's a burrowing toad that spends about half the year hibernating in the loose, sandy soils of East Texas and the other half mating, competing with red imported fire ants for food or getting run over by automobiles. The lethal combination of habitat loss and nighttime encounters with tires is how the Houston toad first wound up on the endangered species list in October 1970. Shortly thereafter, the U.S. Fish and Wildlife Service put Bastrop County on notice that it was one of the last redoubts of the toad and that something had to be done to protect it.

It was years, however, before anyone paid much attention to the toad's plight. Nearly two decades passed and people in Bastrop County carried on with life: They built homes, roads, schools and businesses, turning what was already a fragmented wildlife habitat into mincemeat. With the high-tech capital city of Austin spreading east, the pressure to develop Bastrop County was leading to the loss of breeding ponds, wetlands and grass-tinged pools, the transformation of farmland, and the sometimes careless logging of pine and oak savannahs that once covered the region.

That's when the Fish and Wildlife Service returned to Bastrop County—with the steely-eyed determination of an Old West sheriff.

"The federal government came into Bastrop County [in 1998] and threatened us all with lawsuits," Bob Long said. "They told us if we didn't purchase 15,000 acres of land and set it aside in perpetuity for the recovery of the toad, we would have a huge lawsuit on our hands. Do you know what 15,000 acres would cost the county? Close to $40 million. Our county commissioners said, 'No way!'"

Then again, there was no way the U.S. Fish and Wildlife Service would disappear from the scene. Its agents had an enforcement mandate to pursue. Fearing a litigious impasse, the Texas legislature in 1999 authorized a local task force to come up with a plan that would satisfy the federal agency. The work group assembled all the right people around the table—landowners, environmentalists, utility companies, ranchers and more—but there was still no guarantee that the process wouldn't break down and throw everyone into court.

"Folks were scared. And me, working at the local bank...well, how can we do loans with a lawsuit like this hanging over our heads?" Long asked.

With the support of family, friends and a community network that wanted to save the toad without wrecking the local economy, Long found a better way. In January 2002, he began negotiating for a Safe Harbor agreement that emphasized conservation through sound monitoring and active management over preservation, and incentives over command-and-control.

Safe Harbor agreements are built from the belief that people who do good deeds should not be punished for it. Under such agreements, a landowner commits to doing a "good deed" for endangered wildlife—usually by restoring or enhancing habitats for endangered species—and the government pledges not to "punish" the landowner in the process. The Safe Harbor concept was developed by Environmental Defense, one of the nation's leading conservation groups, and U.S. Fish and Wildlife to encourage private landowners to restore and maintain habitat for endangered species without fear of incurring additional regulatory restrictions.

Since the signing of the first Safe Harbor agreement in North Carolina in 1995, more than two million acres have been enrolled in such agreements and many more are pending or awaiting approval. Landowners engaged in Safe Harbor plans nationwide include private forest owners, residential property owners, land-holding corporations, ranchers, golf courses and even a monastery.

Safe Harbor agreements are setting right some of the perverse disincentives of the Endangered Species Act, which for years has actually given landowners a reason not to come to the aid of rare species. Many landowners fear that if they manage land in ways likely to attract endangered species to their property or increase their numbers, they will suffer more restrictions on the future use of their land.

Safe Harbor agreements effectively freeze a landowner's Endangered Species Act responsibilities at current levels for a particular species if the landowner agrees to restore, enhance or create habitat for that species. This can include the use of prescribed burning in ecosystems that historically were dependent on wildfire disturbance and are now declining because of fire suppression. It can mean longer rotation cycles in forest systems where endangered species are associated with older forest communities. It can mean active control of invasive, non-native grasses and other organisms that threaten ecological integrity. And it can mean re-establishment of hedgerows, vegetated field borders and native vegetation in areas denuded by "clean farming" practices.

In Bastrop County, a combination of practices and partners was needed. Bob Long worked with the Leopold Stewardship Fund partners of Sand County Foundation and Environmental Defense to fence his cattle away from his ponds that had potential to become breeding grounds. Environmental Defense also provided a biologist to listen for and count the toads (the males emit a high, clear trill during mating season). The National Fish and Wildlife Foundation helped with a grant; the Texas Forest Service organized a controlled burn; and the Sand County Foundation provided cash through its Leopold Stewardship Fund.

Long had the responsibility to better manage his herd, including reducing cattle numbers. However, he retained the right to rotate cattle into the breeding area during times of the year when the toad is safely underground and hibernating.

In the spring of 2003, about a year before the Safe Harbor agreement was ceremonially announced on the property, at least ten breeding toads were counted in a fenced pond on Long's ranch. While it cannot be claimed with certainty that the fencing made the pond more inviting to breeding toads, it certainly didn't hurt.

"I'm willing to put my neck on the line and improve the habitat for the toad," Long said. "To me, there isn't a great deal of incentive right away, but I have children and grandchildren. And by me improving the toad habitat that means the duck habitat will be improved. And when the duck habitat improves, so does the turkey habitat. And then come deer and so on.... It's the future that is really worth it. Who knows? Someday I may have property that can be opened up for hunting, something that will benefit my family for generations to come."

Long called the Bastrop County conservation effort an example of "hands-on environmentalism." He explained, "I believe a private landowner can manage land much better than government. The key to this whole thing—and what Aldo Leopold talked about—is that if there are incentives to do this, it is better than government coming in and buying up the land and not managing it."

Safe Harbor agreements negotiated with the help of Environmental Defense (www.environmentaldefense.org) and now other private organizations are helping to save the red-cockaded woodpecker in North Carolina's Sandhills region, the rare northern Aplomado falcon in Texas, and Attwater's prairie chicken in Louisiana and Texas. Safe Harbor agreements aren't appropriate in every situation, of course, nor do they solve every problem faced by landowners whose property is home to endangered species. But a majority of listed species in the United States have their primary residences on private lands. Animals like the rare forest birds in northeastern states don't have the option of moving to secondary locations. There are dozens of species for which private lands can become recovery and enhancement domiciles—if the aware and motivated private landowner can be given legitimate encouragement.

Monitoring, management and Safe Harbor agreements can solve some vexing problems and, in doing so, assure landowners that their continued stewardship won't come back to haunt them. There simply aren't enough public dollars to buy up and manage all the land; private landowners must be entrusted to do what's right.

Bob Long's example is leading the way. Two private properties close to his land with Houston toad habitat have submitted Safe Harbor agreements; more are in the works using the Bob Long Safe Harbor and management template. These landowners are committed to building toad ponds, firing the loblolly pine woodlands, and thinning dense stands of trees that are now inhospitable to Houston toads.

Conservation comes naturally to most landowners. As the story of Bob Long and the nighttime sounds of the Houston toad ring out for the attentive ear to hear, the right incentives can turn a good steward of the land into a great one. And one good steward begets others who also invest their time, their land and their conservation spirit in a species-conserving solution.

19

Where There's Smoke, There's Forest Fire Politics

How community action, science and incentives will restore healthy forests in America

Years of top-down management have turned national forests into matchboxes waiting to be lit. It's time to restore healthier forests by respecting the laws of nature—and getting the incentives more nearly right.

WILDFIRES IN CALIFORNIA BURNED NEARLY 750,000 acres in the fall of 2003, causing 22 deaths, destroying more than 3,600 homes and devastating wildlife habitats and ecosystems. These fires killed millions of trees, fouled the air, sickened people, sterilized the soil for years to come and contributed to runoff that polluted watersheds. Many of those scorched acres eroded severely and caused substantial surface-water degradation during heavy rains in the fall and winter of 2004 and early 2005.

It didn't have to happen.

Sure, the weather in California had been dry, the terrain makes fighting fires treacherous once they start, and people have a tendency to cluster just beyond the shadow of the forest canopy, sometimes courting danger like moths drawn to a flame. But none of this changes the fact that command-and-control government policies, zealously guarded by an environmental crowd that has repeatedly confused preservation with sound forestry management, sparked the fires as surely as an unwatched campfire.

As Congress is belatedly beginning to understand, the time has come to reclaim our national forests and wilderness areas from those who would smother them to death. Forests that look wild and untamed to the backpacking visitor appear sick and even frightening to federal foresters, who know the trees are too crowded, the ground is too brushy and the fire lanes are too inaccessible.

The United States is home to about 297,000 or so square miles (190 million acres) of federal forest and rangeland. That's an area equivalent in size to Texas, or 12 percent of the nation. This incredible resource has always been—and will forever be—susceptible to fire, natural as well as manmade. But the fire patterns of the last decade or more represent a worrying change: there have been more fires that are more devastating, harder to fight and more costly.

Instead of managing our forests wisely, we have allowed wildfires to do the job for us. The fruits of benign neglect are scarred landscapes, obliterated wildlife habitat, streams and rivers clogged with silt, destroyed homes and wasted tax dollars. If an outside enemy had forced this policy upon us, we would call it bioterrorism.

The fires of 2000, 2002 and 2003 were the worst on record. In 2002 alone, 7.2 million acres burned. That's an area larger than Maryland and Rhode Island combined. Arizona, Colorado and Oregon recorded their largest and most destructive wildfires ever. It was also the most expensive wildfire season on record, with suppression efforts costing taxpayers $1.6 billion. Twenty-three firefighters died during the 2002 fire season, tens of thousands of people fled their homes, and 2,000 buildings were destroyed.

These fires were not simply a product of drought, lightning strikes and bad luck. They were the logical result of a century of aggressive fire suppression, attributable in very large part to unconstrained, unbudgeted firefighting expenditures by the U.S. Forest Service in each fire season since 1910, coupled with massive buildups of dense undergrowth, causing forest conditions to deteriorate to an unnatural state.

"After many years of fire suppression, much of America's national forests have tree densities 10 to 20 times natural levels," the Izaak Walton League wrote in its winter 2003 journal. "These heavy fuel loads create potential for catastrophic fires...." Tree stands that once had 100 trees per acre and provided a more diverse habitat are now crowded with 1,000 trees or more, squeezing out wildlife and other species. These conditions have made our forests weaker, more susceptible to disease and insect infestations, and less able to support a healthy wildlife mix. Overcrowding stresses trees, blocks sunlight and reduces water and nutrients. Overcrowding can also turn an ordinary fire into an inferno.

Four factors determine the extent and intensity of forest fires: abundance of fuel, weather, lack of moisture, and terrain. Man has the ability to influence only one of these in a meaningful way: the amount of combustible material in the forest. By reducing available fuel, man can significantly modify the behavior and severity of forest fires.

This is not a problem restricted to the United States: Many of the same ecological circumstances and political-economic constraints face overly fire-prone forests in Australia and Canada. One consequence of the global transport of trees has been a heavy use of eucalyptus from Australia in California. In both the Oakland area fires of the early 1990s and the San Diego and Los Angeles area fires a decade later, many of the houses that burned down were surrounded by rows of exotic eucalyptus, the trunks of which were themselves engulfed in barely decomposed, highly flammable bark, leaves and branches.

Scientifically sound forest-thinning operations—not the mass logging imagined by enviropols—have worked where the private landowners or, more broadly, the U.S. Forest Service has tried them. Many thinned areas survived fires as viable forest habitat while unthinned stretches of forest were turned into scorched wastelands. In the 2002 Cone Fire in California's Lassen National Forest, some 2,000 unthinned acres were reduced to ash and charred trees, but the fire stopped abruptly when it reached a treated area.

"It has been demonstrated that prudent forest management and stewardship can lower the risk of unacceptable loss of property and resource assets through judicious thinning and prescribed burning," said John Helms, professor emeritus of forest resources at the University of California at Berkeley and a member of the Society of American Foresters. "Adaptive, collaborative approaches can lead to sustainable forest management."

An unavoidable truth has emerged: To conserve and nurture a forest, you must sometimes remove a few trees.

The Healthy Forests Initiative proposed by President Bush and approved by Congress in 2003 is a start. It will allow the Forest Service to design a thinning program that will focus on some of those areas at greater risk, perhaps 10 million acres or so. It is the first major forest management legislation in twenty-five years and would streamline approval processes to thin overgrown forests. Features of this initiative, but not the final legislation, had been considered and tentatively agreed upon in negotiations among environmentalists, forest agency leaders, state government representatives, forest industry officials and other stakeholders even before the second President Bush took office.

Federal action alone cannot solve the problem, however. Fuel reduction in forests is a costly proposition. Clearing brush and small trees, which must be done by hand, typically costs more than $500 for a single acre. And not all kinds of forests can be properly treated this way to maintain their integrity. Some forest types, such as lodgepole pine, cannot be sustained by the clearing of brush and small trees. At the rate at which the forests are currently being treated, it would take a century to finish the job. In part, that's because people-hours devoted to thinning must often be redirected to fighting forest fires. A really significant step in greater forest safety will be taken in the United States when funding for forest fire suppression is put on the same budgeted basis as forest planning.

Some of the best opportunities to get incentives right and reduce damage to homes and human lives come in community action and through companies that will manage a person's property to lessen the risk of catastrophic fire. At California's Stevenson Ranch, for example, property owners and civic leaders—unlike those in nearby Scripps Ranch—combined elements of planning and common sense that were successful in preventing any houses there from burning up in the severe Southern California fires of 2003.

From the Swan Valley of Montana, various locations in Colorado, and elsewhere in the western United States comes another means to reduce the human cost of forest fires: the landscape management firm that specializes in controlling trees and other vegetation to keep fires away from buildings and propane tanks. One could imagine an enlightened insurance company giving a cash discount to policyholders who make their homes in the woods safer—and thus make the insurance carrier less susceptible to an expensive claim for property or human losses.

On large expanses of federal land beyond the urbanizing fringe where the great human incursion of housing has taken place in recent decades, what's needed first is a thinning of red tape. That is a goal of the Healthy Forests Initiative, but it must be combined with market-based incentives that will involve private landowners and loggers in the fight to save the forests.

Technology and commitment to expanding jobs—such as what has occurred at the community-based forestry enterprise in Wallowa County, Oregon (see Chapter 21)—can be combined to develop practical uses for small-diameter trees and undergrowth material. If a profitable use can be found for material that is now choking our forests, everyone wins—including the taxpayer, who will pay less to get the job done and will actually benefit from the products.

At the U.S. Forest Products Laboratory in Madison, Wisconsin, researchers are finding ways to:

- Use small-diameter trees in construction.
- Use engineered wood products in wood-frame homebuilding.
- Combine wood fiber with recycled plastics to create composite materials used in windows and doors, signs, roofing, exterior siding and automotive parts.
- Use wood fibers to make inexpensive filters for streams polluted by runoff from mines or farms.
- Use waste-wood chips or sawdust as fuel to generate electricity.

Researchers are also exploring ways to produce ethanol from forest biomass and to improve papermaking so that less water and energy are consumed, and to allow use of mixed wood species.

"Each of those projects could expand the market for small trees and other small forest materials," said Chris Risbrudt, director of the Forest Products Laboratory. "This would encourage ecologically sound forest thinning, reduce the risk of catastrophic fires and make fires less susceptible to insects and disease. It would also help private landowners generate income and resist the temptation to fragment forest areas for development."

A rising demand for forest products from "sustainable" forests can also play a role. Sustainable forestry recognizes that people will always need and want wood—indeed, that's why the U.S. Forest Service was created a century ago. People engaged in sustainable forestry believe that the forests can be better managed through selective cutting of mature trees and removal of the crowded, the crooked and the diseased. It's a logging philosophy of "the worst, first."

The SmartWood certification program is a global example of this new philosophy. Its purpose is to improve the effectiveness of sustainable forestry in conserving biodiversity and providing equity for local communities, fair treatment for workers, and incentives for businesses so they can benefit economically from responsible forestry practices.

Started in 1989, SmartWood is the oldest and most extensive certification program in the world. Initially developed by the Rainforest Alliance to focus on tropical forests, it now works in all forest types—tropical, temperate and boreal—and all kinds of operations, including natural forests, plantations, large commercial enterprises and small-scale community projects. SmartWood has certified more than 800 operations and 25 million acres worldwide, always working with local governments and landowners.

Demand for certified lumber from these operations is increasing rapidly. Products crafted from SmartWood certified wood now include furniture, musical instruments, flooring, and picture and window frames.

The SmartWood approach illustrates the difference between community-based conservation and the kind of trapped-in-time preservation that is one reason why the forests fell into such an unhealthy state. Thus in July 2003, SmartWood and the Rainforest Alliance came out in opposition to the California Heritage Tree Preservation Act (SB 754), which is designed to protect trees on nonfederal forestland that were alive in 1850, marking the year California gained statehood. It also stipulates that heritage trees must meet specific diameter requirements, depending on the species.

While SB 754 may "sound virtuous to anyone who likes a big old redwood or Douglas fir, it would be unfairly punitive to small, private, certified forest owners who are already doing right by their trees," explained Walter Smith of the SmartWood program. The bill would place considerable restraints on the cutting of trees within the heritage tree buffer zones. "If you have a small landowner with twenty acres and a dozen old-growth trees, this bill could end up eliminating a lot of the area that he or she would be allowed to harvest," Smith said. "Our certified landowners are doing a great job of protecting old-growth trees. In fact, foresters are keeping the heritage trees healthy by thinning surrounding trees, which reduces competition for water and nutrients."

Forests cannot be preserved for tomorrow, but they can be managed and conserved. The solutions that will endure and not unduly damage the pocketbooks of taxpayers are those that are built on quality information, a commitment to effective collaboration over the long term, and a stronger role for private landowners and others with a stake in the land.

20

Home on the Range

Western ranchers are fighting development and the barbed wire of environmental red tape

Conservation easements and collaborative management can help western ranchers do what they do best, which is steward their own land.

IN HIS 1939 ESSAY "The Farmer as a Conservationist," Aldo Leopold wrote that conservation "is a positive exercise of skill and insight, not merely a negative exercise of abstinence or caution." He laid out a vision of conservation that called upon private landowners to be active managers, not reactive preservationists. Leopold placed the burden of conservation squarely on the shoulders of those living on the land—the very farmers, ranchers and landowners who each day relied upon the natural world around them.

Successful ranchers embody the proactive Leopold land ethic. They depend on healthy natural resources for their livelihood and way of life; to them, stewardship is not some abstraction to be read in a book but a habit to be practiced every day. Conservation is a cornerstone of survival. Ranchers have learned that environmental practices that conserve and improve the land and water not only make good business sense, they make the *only* business sense. This is especially true as ranches are passed from one generation of stewards to the next.

And yet, ranchers in the western United States face mounting challenges to their ability to maintain their lands and way of life as cities and new development pressures fence them in. From 1982 to 1997, according to the U.S. Department of Agriculture, more than 3.2 million acres of rangeland were developed, mined, quarried or turned into "ranchettes." It is one of the biggest conservation issues confronting the West.

It may also represent one of the largest conservation opportunities. As Leopold wrote some sixty-five years ago, those who are closest to the land and who stake their existence on it are most likely to conserve it.

Pastureland and rangeland represent about 42 percent of land use in the United States, compared with about 46 percent in cropland. More than two-thirds of that land is privately owned. In fact, much of it has been in private hands for a very long time. A 1996 survey by Rockwood Research showed that nearly half of all U.S. cattle businesses have been in the same family for fifty years and 16 percent have been in the same family for seventy-five years.

Keeping the rangelands in private hands is the best way to conserve those lands. The command-and-control preservation edict of public ownership is a losing proposition. Not only does it fail to protect biological diversity, but it's too expensive to be sustained.

More than 70 percent of the nation's fish and wildlife depend on private lands for critical habitat needs. Studies show that most endangered species have most of their habitat on nonfederal lands, and at least 90 percent of endangered species have at least some of their habitat on nonfederal lands. In addition, public lands depend upon interspersed or surrounding private lands for water and habitat needs. Riparian zones are frequently in private ownership. Private lands serve as buffers, migratory corridors, key feeding and breeding grounds and more.

Americans could achieve more fish and wildlife improvement— faster and at lower cost—by working with private landowners. It is a matter of providing incentives for those landowners to do what's best and working locally to develop conservation strategies that work.

Ranchers are helping to lead the way. These examples of private conservation and local partnership were cited by the Sand County Foundation's Leopold Stewardship Fund in 2003:

• The Texas Hill Country Project is restoring habitat for two endangered songbirds, the black-capped vireo and the golden-cheeked warbler, on private land between Austin and San Antonio. Forty-seven landowners who own a total of 101,000 acres are taking part. The Leopold Stewardship Fund provides the services of qualified natural resources consultants. On one property, the number of black-capped vireo nests has increased from fewer than ten to more than forty as a result of habitat improvements and cowbird control efforts.

• The Borderlands Thorn Scrub Project is bringing back habitat on both sides of the Texas-Mexico border to enhance the survival prospects of the ocelot and associated species. Three ranchers with 22,000 acres are taking part. For instance, the Hardy family property is adjacent to a National Wildlife Refuge. The family's full-fledged commitment to stewardship of habitat that benefits the rarest cat in the United States, the beautiful spotted ocelot, makes the refuge a much better place to be a native wild cat. Ecotourism will be a part of this project as a way of making it pay for landowners.

• The Chalk Mountain Project southwest of Fort Worth, Texas, is restoring oak shinnery habitat used by the endangered black-capped vireo. Restoration activities include clearing of juniper, topping of oaks to promote stump sprouting, and prescribed fire. Dr. Rickey Fain, a retired physician from Dallas, has turned his Glen Rose ranch into a guest lodge and conference center that promote ecotourism.

• The Elderberry Longhorn Beetle Restoration Project in California's Central Valley is counteracting the widespread loss of native riparian habitat that has imperiled many species, including the beetle. Rancher Jay Schneider, a board member of the California Cattlemen's Association, has developed a restoration plan and a Safe Harbor agreement with the help of the Sand County Foundation.

• The Utah Prairie Dog Project aims to produce agreements among ranchers to restore rangeland in a way that would benefit livestock as well as the threatened Utah prairie dogs. It will shield those ranchers from Endangered Species Act regulations through Safe Harbor agreements (see Chapter 18).

• The Montana Water Trust Project is buying or leasing senior water rights in northwestern Montana's Dayton Creek. The trust will compensate landowners and ranchers for the value of water rights needed to sustain rare fish, such as bull trout and west slope cutthroat trout, which need more water in the stream for migration and spawning.

• The Owyhee Sage Grouse Local Working Group is conducting aerial surveys in southwestern Idaho to document leks (breeding ground) and count birds in those leks in the Big Jack's Creek and Diskshooter Ridge areas. Increasing the information on lek locations and sage grouse populations will allow comparisons with historic data and enable better assessments over time. The Leopold Stewardship Fund is helping establish resource banks, similar to grass banks, and remove juniper to enhance sage grouse habitat of sage brush and understory vegetation. Cattle rancher John Romero, the chairman of the Idaho Cattle Association Wildlife Committee, is co-chairman of this project.

A quiet revolution is growing among American ranchers. It has no single modern leader or manifesto, and its expressions are as different as the ranches and rangelands where it has emerged. But at the core of this revolution are the words of Aldo Leopold and others who practice hands-on environmentalism.

The efforts of the Colorado Cattlemen's Agricultural Land Trust (www.ccalt.org) reflect the Leopold land ethic as well as any example in the American West. It is a story that illustrates the power of people acting in unison to conserve land and a way of life, and the capability of ranchers who do the work in their own unique ways. Recognizing the need to help Colorado ranchers and farmers protect their lands in the face of mounting development and economic pressures, the Colorado Cattlemen's Association founded the land trust in 1995. It has helped ranchers put about 132,000 acres of working cattle ranches under "conservation easement," a tool that secures conservation value on private land while conveying some economic benefit to the landowner or his successors.

One such rancher was Sam Capp, whose family has ranched in Colorado's Huerfano County since 1872, when his grandfather migrated from England and settled near the Spanish Peaks between Pueblo and Trinidad. The land was lost during the Depression, but a young Sam Capp and his father bought back 5,000 acres, which they ranched with just two horses. Today, the Capp Ranch has grown to 28,000 acres; it has been conveyed to the stewardship of Frankie and Sue Menegatti after they spent years helping Sam Capp make the ranch a conservation success; and it's as healthy as any in Colorado.

And that's precisely the problem. Colorado is losing tens of thousands of acres of productive land each year, changing the land-scape in ways that aren't usually compatible with those farms and ranches that remain. It also breaks up and isolates wildlife habitat and strains the use of one of the West's most precious resources—water.

Sam Capp and his successors, Frankie and Sue Menegatti, wanted to sustain the ranching way of life for future generations. They also wanted to conserve the land and the wildlife. But today's economic currents make that difficult without a plan. For Capp and the Menegattis, that plan was a conservation easement nego-tiated with the help of the Colorado Cattlemen's Agricultural Land Trust.

A conservation easement is a legally recorded agreement between a landowner and a qualified conservation organization that restricts land to specified uses, such as agriculture, watershed protection or open space. In placing a conservation easement on their property, landowners voluntarily limit their ability to develop their property, thereby permanently protecting its open space and agricultural values, without allowing public access to the protected property unless that is the specific wish of a particular landowner. An agricultural easement generally prohibits or limits the right to develop the land for nonagricultural uses. These rights are then extinguished and cannot be sold or transferred to another entity.

By donating these relinquished rights and by meeting specific conditions, a landowner may become eligible for certain tax benefits. The organization receiving the easement accepts the restrictions and seeks an endowment to cover the costs of those long-term obligations. Federal tax benefits and state tax credits are available only for perpetual easements that subject all future landowners to the easement restrictions.

Donating developing rights through a conservation easement:

- Generally prohibits or limits any subdivision, development or any practice that would damage the agricultural value or productivity of the land.
- Ensures that these conserved landscapes will remain available for agriculture for future generations.
- Significantly reduces federal and state income taxes, estate and inheritance taxes.
- Qualifies taxpayers for an income tax credit in participating states. These credits are able to offset state income tax and can be sold to another taxpayer, helping a landowner realize cash benefits for the donation of a conservation easement.

Conservation easements can be placed on mortgaged property, and the presence of a conservation easement does not eliminate the possibility of securing future loans with that property. In the case of the Capps' ranch, the easement states that no open mining or gravel pits can be created on the 28,000 acres. It also sets a predetermined number of buildings, which prevents development.

"Basically, it says there shall be no disturbance of the land that doesn't benefit livestock," the late Sam Capp said in *The Roundup,* a publication of the land trust. "All my hard work will be preserved. It won't be wasted on development.... This is something anyone with a ranch should consider.... This conservation easement will preserve open space, game habitat and the ranching and farming way of life."

The conservation easement alone is not a conservation strategy, however. Sam Capp and the Menegatti family have taken many other steps over time, including cross-fencing to reduce rotation, moving cattle away from riparian areas to ridge tops and side hills, better care of food plots, small-diameter logging and selective cutting, noxious weed control and other habitat improvements. The result is a ranch that works better for man and nature.

American ranchers are the front line of defense in conserving the natural West. They are doing so, often quietly, usually over the uninformed tirades of city-born environmentalists who wouldn't know a calf from a rope, because they love the land and their way of life. Stewards acting privately and collectively are making a difference on the American ranch by exercising "skill and insight," as Leopold counseled, not "abstinence or caution."

21
Starting Up

A blueprint for getting your hands around an environmental opportunity in your community or state

Being an effective twenty-first-century environmentalist won't mean chaining yourself to a tree or worshipping at the altar of regulatory control. The new "hands-on" environmentalist will think locally and act locally, usually in cooperation with partners who share values and a commitment to a community-based process. Here's how to get started down a more effective and more gratifying path.

GARRETT HARDIN MAY HAVE BEEN BEST KNOWN for his 1968 article in *Science* magazine, in which he coined the term "tragedy of the commons" to describe the human tendency to destroy a shared or "common" resource, such as ocean fisheries, absent private ownership or public regulation. The phrase became part of the ideological cant for environmentalists of that era, including some who ignored Hardin's nod to private ownership in their rush to embrace top-down regulation.

Here's another idea from the late Professor Hardin that should become a mantra for the evolving civic environmentalism of the twenty-first century: "Never globalize a problem if it can possibly be dealt with locally." Hardin first wrote this in his 1985 book, *Filters against Folly: How to Survive Despite Ecologists, Economists and the Merely Eloquent.* More than ten years later, he illustrated what he meant in an interview with *Sceptic* magazine.

"We could form a global committee to fill in the potholes of the world and the result would be a disaster," Hardin told interviewer Frank Meile. "Imagine if you had a pothole in front of your house and you couldn't get it filled in until you got approval from some council thousands of miles away representing 6 billion people. Don't be silly. Fill in the pothole yourself."

Garrett Hardin was no lockstep conservative. In fact, he pushed for world population control since the early 1960s and often warned that market economies don't function properly if environmental costs are spread over the larger community while profits are privatized. He understood, however, that people should not call on the United Nations or the Environmental Protection Agency to "fill" an environmental pothole. That's a job best done close to home, by people who are willing to get their hands dirty shoveling a little asphalt.

If you've never before filled a pothole, it can be difficult to know how to get started. Where do I buy the asphalt? What patching material will bond with what's already on the road? What's best for my climate and the amount of traffic on my road? How much will materials cost? Can I really do this myself, or do I need help? How do I go about getting a government permit?

Citizens contemplating environmental challenges in their neighborhood, city, region or state may ask themselves very similar questions about what to do first, second...and beyond. The first thing they will discover, however, is there's no single textbook for organizing a community-based environmental effort. Logically, there can't be. Community-based environmentalism is all about fashioning local solutions to local problems using local citizens. While command-and-control environmentalism relies on regulation, process and conformity, hands-on environmentalism is rooted in creativity, adaptability, improvisation and a fair bit of spontaneity. The most effective community-based projects are unique to the place and the social context in which they are carried out.

Like any undertaking, a community-based environmental project must be organized in order to work. It must have a beginning, a middle and an end that often loops back to a new beginning. Here's one simple model that has worked for groups large and small, established and new:

1. First, assess the situation.
2. Next, plan your approach.

3. Then, execute your plan.

4. Finally, evaluate results and make adjustments where necessary.

Think of these as four basic steps for getting your two hands around an environmental opportunity in your community. Or, think of them as a blueprint for building a virtual house. The Greek word for house is *oikos,* and modern ecology is the study of the "house" we call nature. Here's a blueprint with examples of how some groups got things done.

Step One: Assess the Situation

ALDO LEOPOLD WAS THE FIRST TO articulate a land ethic—an obligation to respect the rights of the land and to protect its health by nurturing its biotic processes. In short, Leopold urged people to make a nature connection, to understand how human interaction with the land can, and should, be for the benefit of both.

"Who is the land?" Leopold asked. "We are, but no less the meanest flower that blows. Land ecology at the outset discards the fallacious notion that the wild community is one thing, the human community another."

Making a nature connection can be as simple as deciding that a local stream or creek should be cleaned, an aging dam removed, a small prairie saved, a nesting ground for sea turtles preserved, or a long-neglected piece of land restored to a more natural state. That connection begins with a sensory challenge that is intensely local, one that can be seen with the eyes, smelled with the nose or felt with the hands.

"It all really begins with citizens who have an informed sense of place," said Curt Meine, a wildlife biologist who helped create a unique community effort to protect and restore the 7,354 acres surrounding an obsolete Army ammunition plant in south-central Wisconsin. "It begins with a curiosity about the soils, waters, plants and animals of the region; appreciation of the connections between natural history and human history; a loyalty to local landscapes and communities; a desire to overcome entrenched political and cultural divisions; a willingness to listen and share information, and to engage in honest conversations. Without these, you don't get to first base. You're sitting in the dugout."

That connection may often begin by tapping into the passion of a friend or neighbor. Rarely do people who aren't linked to the land rise off their sofa and announce, "See you later, Honey. I'm going out to clean up Beaver Creek." More often, the uninitiated are introduced to an opportunity by someone who has already assessed what's at stake.

Also, citizens don't so often "pick" an environmental issue as have one thrust on them. There are challenges and opportunities in virtually every community, but significant efforts usually start with some sort of catalyst. There may be a sense of urgency or crisis or an impending public or private action. In other words, you become aware that something is about to happen (or should happen) in your own back yard.

Even then, life can get in the way. "Most efforts don't 'get started' because conscientious citizens just feel too alone, or too powerless, or too busy, or too tired to do anything about it," said Meine, who co-edited *The Essential Leopold: Quotations and Commentaries*. Unless there's an initial commitment by citizens to "do something," most good intentions die for lack of attention.

Or, sometimes, fear of the unknown. "We all want progress but we're terrified of change," said Todd Ambs, formerly of the River Alliance of Wisconsin and now in charge of water issues for the Wisconsin Department of Natural Resources. "It's important to realize how resistant we humans can be to any kind of change in our routine."

How does one choose from among the many environmental issues that are likely to surface in any community? Start by picking one that has a fighting chance to be resolved. That doesn't mean picking a "slam-dunk" issue that could be resolved by just about any group. After all, if hands-on environmentalism were easy, everyone would be doing it. In fact, it's often the most vexing environmental issues—those made worse by command-and-control approaches or a history of polarized politics—that most need and demand a fresh, community-based approach.

Be prepared to learn that most issues are more complicated than they appear. One challenge may be connected to another. A watershed group, for example, may find itself working on intertwined issues within that watershed.

Get the "facts." Then check them. The "facts" as presented in local news stories (which may be overly confrontational) or by special-interest groups may not hold up under scrutiny. Successful community-based conservation groups find a way to tap into reliable science, perhaps with help from a college or university, a company with an environmental engineering department, or a state natural resources agency. Consensus can be built around shared, reliable information.

The worldwide campaign to block the use of DDT, an inexpensive chemical that kills malaria-bearing mosquitoes with minor effects on the environment, offers a classic example of how partially understood "facts" can produce wrongheaded results.

Developed during World War II, DDT (short for dichloro-diphenyl-trichloroethane) was at first viewed as a lifesaver. In Asia, Africa and Latin America, where DDT was used to control mosquito populations, malaria was on its way to being eradicated by the 1960s. But the abuse of DDT in the United States—primarily by federal agencies that insisted on massive, generalized sprayings—prompted protests and lawsuits by farmers who rightfully objected to the chemical carpet-bombing of their property. Overused or sprayed indiscriminately from the air, DDT can kill fish, birds or their eggs. Used properly, however, it can save millions of lives without harming humans or the broader environment.

"Malaria kills 1 million people a year, mostly children, and sickens hundreds of millions," wrote Roger E. Meiners and Andrew P. Morris in "Pesticides and Property Rights," a report for the Political Economy Research Center (PERC) in Bozeman, Montana.

> Fortunately, a cheap preventive measure is available. An inexpensive generic chemical can be sprayed on the walls of residences.... The chemical is effective on mosquitoes and some other insects, and the evidence from decades of use is that, unless abused, it has no ill effects on humans and a minor impact on the environment other than disease-carrying insects. Applied once every six months, it can greatly reduce the mosquito problem. Millions of lives can be saved and hundreds of millions will suffer less. Unfortunately, the chemical is DDT.

DDT began its slide into political disfavor in the late 1950s, when organic farmers on Long Island filed suit to prevent spraying over their land. Then came the publication of *Silent Spring*, the 1962 book by Rachel Carson that became an anthem for the environmental movement. Backed by evidence that aerial spraying of DDT could kill birds and fish, Carson warned of a spring in which no birds would sing because they would have been killed by constant exposure to the chemical. There was an element of hype in Carson's apocalyptic prediction, but pressure nonetheless mounted for DDT's ban.

Like an alcoholic who suddenly becomes a self-righteous teetotaler, the government went from being a chronic abuser of DDT to prohibiting its use in 1972 and eventually halting its domestic production. True to command-and-control style, the government's rules left no room for careful applications of DDT under special circumstances or for periodic review of the science. Virtually overnight, DDT was transformed from a public good to a public evil.

Inevitably, malaria began its deadly comeback. "The disease is nearly back to where it was 50 years ago," wrote Meiners and Morris, noting that only three nations produce DDT today and only twenty-three use it. And yet, malaria kills a child somewhere in the developing world every thirty seconds.

Other chemicals and programs have proved far less effective in controlling mosquitoes, which is why some scientists—including three Nobel laureates in medicine—have called for spraying DDT inside houses in places where people want it. The science supports that careful approach, but the politics don't. Environmental groups aren't about to back off one of their biggest and most identifiable victories, the virtual eradication of DDT. "Since the early 1970s, opponents of DDT have held the upper hand politically," wrote Meiners and Morris. "Political action, unlike markets and the rule of law, tends to dictate one solution for all."

The facts about DDT, having been compromised decades ago, may never again be a factor in the debate. Because one bad centralized planning decision (massive spraying) was replaced with another (a ban on DDT), a cheap, targeted and readily available solution to the ravages of malaria will likely remain in political exile.

The DDT story might have ended differently had all sides

agreed on what facts were not known. In emotional or politicized situations, opposing sides may have trouble finding common ground beyond their mutual disdain. Sometimes, polarized groups can agree to work together to gather basic scientific data—a step that often begins the essential process of building trust.

Careful assessment also includes sizing up the field. There may be existing groups trying to do what you want to do, or a stalled effort that needs revitalization. Identifying or listing critical stakeholders is another important housekeeping item. Meeting with landholders or resource owners is a logical outgrowth of such a list. It's also important to communicate with other decision makers who may have influence, even if it's indirect.

By now, you probably have a pretty good idea of what you're getting yourself into. But the real work (and some fun) lies ahead.

Step Two: Plan Your Approach

ALDO LEOPOLD WAS CONVINCED THAT people, planning and then acting as individuals or a group, could do right by the land. Although he lived before "It's not rocket science" was a part of the language, Leopold made it clear that doing something to help the land need not be as complicated as sending a man to the moon. In *A Sand County Almanac* he explained:

> Acts of creation are ordinarily reserved for gods and poets, but humbler folk may circumvent this restriction if they know how. To plant a pine, for example, one need be neither god nor poet; one need only own a good shovel. By virtue of this curious loophole in the rules, any clodhopper may say: Let there be a tree—and there will be one.

Some level of planning is essential to the success of most environmental undertakings. If you're acting alone on your own property, it can be as simple as picking the right spot to plant the pine tree and making sure your shovel is sharp. If you're working with others, the essence of the community-based approach to conservation, other factors will come into play.

In the assessment stage, you discover who your potential partners may be. You collect and check the facts. You begin communicating within the larger circle. Planning your approach involves thoughtful integration of all three.

Building a lasting, community-based partnership may begin with "setting a table" around which everyone can meet. Taken literally, this means finding a physical place to meet that doesn't put one group or another on the defensive. Beyond that, it means creating an environment that is as inviting and neutral as possible.

Some questions to consider: What are the rules of conduct at a meeting? What values do various partners bring to the table? Do they have incentives—financial, cultural or environmental—to work together? Can the group agree on a common set of challenges or issues? If not, can they at least agree on a sifting and winnowing process to narrow the list?

Part of planning is writing a "mission statement" that can serve as a touchstone for everyone involved. A good mission statement should capture your organization's reason for being. "It is the single, consistent message that you, your board members, other volunteers and eventually your staff will use to gather support and attract funders," advised the River Network in its comprehensive guide, *Starting Up: A Handbook for New River and Watershed Organizations*. Don't expect to write a mission statement in one meeting, the River Network cautions. Use a couple of meetings to build consensus.

Re-evaluating the facts is part of the planning process, too. Your new partners may bring perspectives to the "facts"—scientific, political, cultural or social—that escaped you at first glance. Reassessing the facts leads to building a sound information baseline, which becomes a way of charting where you're going and monitoring how well you're doing. Facts set parameters of what's known, but careful and sustainable monitoring rests on checking only the most important variables. Unless the facts produce a plan that is meaningful and relevant to all involved, they're nothing more than numbers and words. In regard to communities coming together to support floodplain rehabilitation through removal of unsafe dams, a growing matter of concern across the United States, there are publications about essential facts by the Heinz Center. Under the heading of "Dam Removal Research" these documents help guide conservation practitioners to the essential facts they will need to improve their rivers.

A good plan today may be better than a perfect plan tomorrow or the next day. The four steps—assessment, planning, execution and evaluation—rarely take place in a neat sequence. Sometimes it's necessary to get off and running before the full "plan," in its most formal sense, is completed. The best plan may indeed be one that everyone at the table describes as a "no-brainer," meaning it's plainly evident how each group or person wins.

Planning can be especially important when problems seem most intractable or they're rooted in a history of mistakes and mistrust. The planning to build a sustainable environment and economy in the Navajo "New Lands" area of Arizona illustrates this point.

The New Lands area represents a fresh start for many Navajo ranchers, who until the mid 1990s were living on parched lands that had been exhausted, in some cases, by their own grazing practices. Navajo ranchers don't want to repeat the mistakes of the past, but they've also got good reason not to trust outsiders who prescribe quick remedies. There are still many Navajo ranchers who remember the Bureau of Indian Affairs solution to overgrazing in the 1930s: Slaughter hundreds of thousands of Navajo sheep and cattle and dump them in mass graves.

To the Navajo, ranching cattle and sheep is a part of their connection to the land. In the tribe's creation stories, Navajos are said to have domesticated animals such as jackrabbits, beavers and turkeys, while ownership of those animals remained with the "holy people," the ancient gods.

Even today, livestock can be more important to the Navajo than money. Yet many Navajo ranchers recognize that overgrazing is bad for the land and, ultimately, for them. Droughts in the 1990s added a sense of urgency to the question: Can Navajo traditions live within a new and more sustainable ranching ethic?

The search for answers in the New Lands area is tied to the Dineh Bi' Ranchers Roundtable and Development Corporation and its "Help Our Mother Earth" project, which is essentially an effort to integrate a new land ethic with the old. The project, with help from groups such as the Sand County Foundation, aims to create a community-based problem-solving model that meshes with Navajo customs.

Fortunately, the Dineh (Navajo) Bi' Roundtable enjoys a measure of autonomy from a tribal council that can be just as guilty of one-size-fits-all thinking as any government agency. The roundtable has three goals: Monthly educational meetings for interested ranchers, creation of a rancher "certification" program, and training to help make ranchers and other range bosses better monitors and stewards. The certification program includes training in water facility and fence maintenance, cattle production and record keeping, grazing pattern planning, plant identification, and better understanding of range ecosystems. By monitoring vegetation, soils, animals and weather patterns, the Dineh Bi' ranchers will make better grazing management systems and help avoid future "tragedies of the commons."

"Our people, who have little experience with rule-of-law systems of planning and regulations, have a real need to be included and coached in participatory representative government in order to achieve the economic and community goals they have, while sustaining the ecological integrity of the land," wrote LeRoy Begay, a rancher who is president of the Dineh Bi' group. "Current federal (BIA) and tribal programs do not work in attaining goals, as their policy framework is 'top down' and does not build community (habitat) level management, which is essential for effective communal land management."

It's too early to tell whether the Dineh Bi' Ranchers Roundtable will work, but one thing is known: Without planning tailored to the Navajo ranchers, their land and their history, the New Lands would still be ranched today in all the old ways.

Step Three: Execute Your Plan

YOU'VE ASSESSED THE CHALLENGES and opportunities. You've "set a table" for collaborative discussion. You're ready and eager to get to work....

Not so fast. Depending on the nature of your group, there may still be a few nitty-gritty details awaiting you.

Pick a name. You don't have thousands of dollars to hire an image consultant, but you want to come up with a name that captures a mission, an image and a sense of place for your group. "Ideally, it should be positive, descriptive and simple," wrote David Bolling in the River Network's *Starting Up* guide. "It should be for something, not against something."

Don't be cutesy—but don't be boring, either. And remember that your opponents might make hay with the wrong name. If you're the South-Central Area Mothers for the Environment, don't be surprised if the acronym "SCAM Environment" gets tossed around.

File articles of incorporation. It's not always necessary for a nonprofit group to incorporate, but it makes things a lot easier if you're hoping to sustain the work of the group over time. Incorporating lends valuable credibility and structure to your effort, provides limited liability protection for board members and makes it possible to file for tax-exempt status. Groups can incorporate with the help of an attorney in many states, but it's usually best to get advice. Maybe you're lucky enough to have a lawyer in your group who's willing to offer services *pro bono* (free of charge), or perhaps you can find volunteer or low-cost legal help through other contacts.

Here's the bottom line: If you want to raise money for your cause, and if you want your donors to be able to take a tax deduction for their donations, incorporation is a necessary first step.

Draft bylaws. If your group incorporates, bylaws are necessary under the law. But they're more than just some extra paperwork to satisfy federal and state requirements. They're a roadmap for how your group will operate.

This is where your mission statement gets expanded into language describing how the group will be organized, led and managed. There are "model" bylaws available and many lawyers can adapt boilerplate language to fit your organization's needs. No matter how they're written, bylaws should be formally adopted during the first official meeting of the board of directors following incorporation. Don't worry; they can always be amended later if something doesn't work out.

File for tax-exempt status. Becoming a 501c3 organization under the Internal Revenue Code isn't for everyone. It creates an opportunity for your donors to get tax deductions and it exempts your organization from paying taxes on its income, but it comes with a lot of rules and reporting requirements attached. Still, many community-based groups wind up filing for tax-exempt status if one of their goals is to acquire and maintain property.

"Once you have a plan in mind, filing for tax-exempt status is one of the first things you do," said Randy Creeger, a financial adviser who has helped establish the Touch the Sky Prairie Refuge in Luverne, Minnesota. "Find someone with experience to handle that for you. It's that important."

Again, these steps are not necessary for every group. Many hands-on environmental initiatives function nicely without legal bells and whistles because their mission doesn't require them. If the challenge is large enough, however, a structured organization may be necessary.

One of the most successful nonprofit, hands-on endeavors is Wallowa Resources in Enterprise, Oregon (www.wallowaresources.org). The group has done nothing less than bring back sustainable logging in the rural Northwest by balancing economic and environmental needs.

In 1993, the Chinook salmon landed on the federal endangered species list. That prompted stiff logging limits in and around the Wallowa-Whitman National Forest, a breeding ground for the Chinook, and about four hundred loggers lost their jobs. Wallowa County's three mills shut down and unemployment soared to 15 percent. Tempers shot up, too. Someone tried to burn down the local office of the U.S. Forest Service and two local environmentalists were hanged in effigy.

That's when Sustainable Northwest entered the picture. This nonprofit based in Portland, Oregon, helped launch Wallowa Resources with seed money and a vision for a greener local economy. Clearing brush and smaller trees has become a way to help bring back the once-thick forests, eliminate fire hazards and make money. Some one hundred logging or related "green" jobs have been restored. They're no replacement, yet, for the jobs that were lost. But the comeback is heartening for a community that had given up hope.

"The organization emphasizes community empowerment, in balance with state and federal agency cooperation," reads a project description. "Through demonstration projects, educational programs, and market-based incentives—essential tools—Wallowa Resources is working to create a healthy environment with a new, healthier rural economy."

What may be most unusual about Wallowa Resources is that it has taken a financial risk. In March 2001, Wallowa Resources invested about $100,000 to reopen the Joseph Timber Company mill after the mill's other sources of funding were tapped out. The nonprofit took a 10 percent equity position and 50 percent management control of the mill. As of late 2001, the mill was selling timber from new-growth trees to a home developer in Bend, Oregon.

Earlier, Wallowa Resources rented a portable sawmill to cut odd lengths of lumber out of fallen and unhealthy trees. But there wasn't a market for the four-by-fours, and the group only recouped its $24,000 investment by selling the lumber for wood chips.

Whether or not every venture is successful, Wallowa Resources is proving itself to the community by becoming a decision maker with a stake in the outcome. Executing a plan this ambitious wouldn't be possible without the right structure—and the right commitment to a new and more entrepreneurial environmentalism.

Step Four: Evaluate Results

FOUNDED FIFTY YEARS AGO, THE Nature Conservancy is one of the world's most successful environmental groups. Through private and public partnerships, it has preserved more than 92 million acres around the world, from native prairies to tropical rainforests. Each Nature Conservancy project is different, of course, but the management and decision-making practices employed are remarkably consistent.

The Nature Conservancy's "Conservation by Design" model has four main components: Setting priorities, developing strategies, taking action and measuring success. It's another version of "adaptive management," a cycle of continuous improvement based on learning from the outcomes of operational programs. The lessons learned by evaluating successes and failures are used to set fresh priorities, which leads to revised strategies and new actions. This is the cycle of organizational sustainability for many hands-on environmental groups.

Evaluation usually begins with measuring interim results while a project is under way, rather than waiting until it's too late to correct course. It requires reporting those results to partners, soliciting advice, adjusting the work and sometimes amending the budget or the plan itself. Environmental groups need not feel embarrassed, frustrated or unsuccessful if their first pass at a challenge doesn't work out as planned. Just ask the people of the town of Amherst, near Buffalo, New York.

Amherst is a textbook example of what happens when suburban growth changes a biotic community. The population of white-tailed deer in Amherst seemed to grow with the human population, in large part because there was still ample food in suburban gardens and a ban on hunting. The result was a sharp rise in deer/vehicle accidents—about five hundred reported cases in 1996 alone, or 10 percent of all motor vehicle accidents investigated by local police. The highways were less safe, the paint and body shops were busier and the insurance companies (and policyholders) were poorer.

Local officials checked the pulse of the community and decided in the mid 1990s that "lethal control," or shooting, of deer was the answer. It worked in one sense: deer/vehicle accidents declined. But Bambi soon ran headlong into NIMBY. Shooting became unpopular and was discontinued after three years, leaving Amherst pretty much where it started: with too many deer running into too many cars.

What emerged was a more integrated approach that involved more and better record keeping, monitoring and efforts to change the behavior of people as much as the deer. Nuisance deer control continues at a modest scale, but only on "heritage farms" that will not be developed and will be preserved as a reminder of the town's historic roots. The town has built deer/vehicle accident statistics (about 3,300 reported accidents and a similar number of carcass pickups) into its geographical information system. The information is being used to develop new strategies with more public input, which, in turn, could lead to more sustainable solutions.

The final word on deer management plans in Amherst and

other suburban communities with deer problems may be years away. Meanwhile, other efforts to better manage North America's deer population continue.

Through the Sand County Foundation and other hands-on environmental groups, "Quality Hunting Ecology" demonstration areas have been established on more than 100,000 acres in the Great Lakes states and the Northeast. Landowners, sportsmen and resource agencies are working together to improve habitat health and the quality of recreational opportunities.

The cost of too many deer is hard to deny. There are about 1.5 million deer/vehicle collisions each year, with thousands of injuries and many deaths. In 1994 alone, there were 211 deaths and $1 billion in insurance claims. A decade later, both figures were much higher, particularly the economic losses. Deer also spread Lyme disease to humans, devour crops and thwart sustainable forestry's aim for decent timber production by eating their way through young, healthy trees.

One of the most successful Quality Hunting Ecology programs so far is the "Earn Your Buck" initiative. In this program, hunters agree to kill two doe before shooting a buck. The goal is to improve the health of the herd while allowing the habitat to recover from overbrowsing.

Monitoring and evaluation will be the key to the success of the Quality Hunting Ecology program. Research into landowners' and hunters' attitudes must continue, baseline data on deer and habitat must be established, and results must be measured in five participating sites.

"Evaluation and monitoring go hand in hand," wrote Will Allen, who developed an Internet site on emerging conservation strategies for the Natural Resource Management Programme of Massey University in New Zealand (www.nrm-changelinks.net).

> Monitoring provides the raw data to answer questions. But in and of itself, it is a useless and expensive exercise.... Monitoring for monitoring's sake is monitoring that should never be done.
>
> Evaluation is putting those data to use and thus giving them value. Evaluation is where the learning occurs, questions answered, recommendations made, and improvements suggested. Yet without monitoring, evaluation would have no

foundation, have no raw material to work with, and be limited to the realm of speculation. As the old song says, "You can't have one without the other."

It may be an old song, but it's a new tune for a brand of twenty-first-century environmentalism based on science, voluntary public participation, and market-based solutions.

22
Thundering Back from the Brink
Drawing lessons from the past to solve today's
problems

*Hands-on environmentalism works from sound principles. When these are
deployed, over the long term, by imaginative landowners and their partners, they
ensure improved land health and greater well-being of people on the land. In
contrast, top-down, mandated environmentalism works against, not with, the better features of human nature and is producing fewer and fewer results at
ever-greater costs. The environmental future is bright when and where people
are inspired, not coerced.*

IF SACAGAWEA, INTREPID INTERPRETER and the only woman on the
Lewis and Clark Expedition, had been able to live out a full three-
score-and-ten-year lifespan, her remarkable life would have
encompassed much of a century in which bison, a major food for
the expedition, were nearly eliminated from the American West.
In contrast, an elderly woman venturing now to vacation where
Sacagawea labored would witness what might have been unimaginable at the time she was a young woman herself in the early
twentieth century—the flourishing of bison on many of the lands
that had been killing fields for wildlife in the 1800s. At the start of
the 1900s, bison were thought to have a future only in zoos.

Two centuries have come and gone since the start of the Lewis
and Clark Expedition. Massive transformation of wildlife communities and habitat has occurred across most of the North American
continent. But bison are back. In the tens of thousands they live
on private land, and in dozens or hundreds they are tended in public parks and refuges. Continued existence of this magnificent
prairie grazer is not in doubt.

The ponderous symbol of the U.S. and Canadian western
plains owes its survival and resurgence to a classic case of hands-on

environmentalism. Several private owners of bison kept some small herds alive and secure. Thus it was possible for the charismatic William Hornaday, a hunter for the museum trade who became a preservationist, to push and cajole private and public collaborators to prevent bison from being extirpated in the United States and later to re-establish whole herds.

Ironically, the same William Hornaday who championed the Bronx Zoo and the establishment of bison herds in different places caused twenty-five of the last free-ranging bison in the United States to be gunned down in 1886. Hornaday chose among them to secure half a dozen specimens to mount and display in the U.S. National Museum. Three of the six, he later wrote, "were killed by yours truly." Fortunately for those who hold wildlife dear, while Hornaday was plundering bison on the commons of the northern Great Plains, private owners were becoming stewards to bison that were protected behind barbed wire and fence posts.

Establishing the possibility of benefit, such as securing ownership of bison, is an essential precursor to sustainable responsibility. Without people owning bison in the late 1800s, the recovery might never have developed; without the possibility of ownership, stewardship opportunities and economic benefits currently associated with bison would be restricted to small numbers on isolated lands held by government.

Today, the situation for wildlife in the southern African nation of Zimbabwe is similar to that faced by bison and other game of North America at the beginning of the twentieth century (although the conditions for people are considerably worse). Both the bison and the black rhino were nearly eliminated from public lands; both were being kept in zoos at considerable expense and in low numbers; and both species were also being protected by landowners. The question is now being asked, "Will the black rhino survive?" If the answer is going to be "yes" decades hence, it will be accomplished by a convergence of the same types of interests and motivations that brought bison in the United States back from the brink and into a promising future.

As we have shown, and as has been documented in the *New York Times* and other media, conservancies in Zimbabwe hold black rhinos more securely and in greater numbers than do the national parks of that troubled nation. Even if the despotic Robert Mugabe and his cronies continue to tighten their hold on owners of the conservancies such as Bubiana and wildlife suffers further, too, we have witnessed enough to know that once again a vital spark of wildlife conservation is being kept alive by landowners. There may be a future for the black rhino yet in Zimbabwe.

But what is the future for fish in the sea? Although seeming to be an inexhaustible resource during most of world history, many of the world's ocean fishes are now severely depleted. There is considerable doubt among experts that a number of once-flourishing fisheries can ever recover. For instance, the Georges Bank cod fishery of eastern North America may be economically extinct. To bring back the vitality and the economic value of the blue-water realm, it is necessary to identify that combination of interests and passionate motivation that brought back the bison and is keeping hope alive for black rhinos. We need to put a similar combination to work in behalf of shellfish, coral reef animals, fish of the open ocean, and the people who rely on these creatures in their habitats for food and livelihood.

Like bison on the open range, which could be killed or taken at the cost of labor and materials, many ocean species are taken without immediate negative consequence to the taker. One glaring example that results in white, dead heaps of coral—with a macabre analogy to the bone-and-skull heaps from bison carcasses on the Great Plains—is the poisoning and blasting done increasingly to extract marketable fish from tropical coral reefs. Even coral reef areas within designated parks are assaulted. Whether to enrich a distant fish fancier's aquarium or to hold a temporary place in a fishmonger's stall before being purchased and eaten, fish taken this way are taken destructively. The capacity of habitat to recover is severely impaired.

Unyielding but spectacularly unsuccessful government agencies are trying to keep full control of marine biological resources. Despite their good intentions and the hard work of public officials, they rarely succeed in maintaining populations of economically important protein sources for their own citizens. They cannot create enough rules to stop the use of explosives or poisons. But even these difficult circumstances may yield to creativity, persistence and need.

Because there are not enough incorruptible government officials in any nation to protect resources on the "open range," nor enough money to compensate government rangers adequately, people with conserving and humanitarian interests both at heart have begun to explore sustainable alternatives to heavily policed parks. Some of the more promising solutions build from the ground up. People who are poor but live close to quality natural resources may develop a sense of ownership if meaningful benefits start to flow their way.

In Indonesia, parts of the Bunaken National Park have become the responsibility of a new local advisory board. The North Sulawesi Tourism Promotion Board explains the solution being built from the ocean floor on up:

> Bunaken's entrance fee system is the first of its kind in Asia, and is being held up as a model system by marine conservationists around the world. The most important aspect of Bunaken's system is that the money collected remains with the Bunaken Management Advisory Board to fund conservation and village development programs in the park—instead of heading to the national coffers as with every other national park in Asia (and many throughout the world)! This makes a world of difference, as it means your money goes towards managing the very reefs you've come to enjoy.

The incentives are being arranged to secure more local involvement and, hopefully, a great deal of local citizens' control:

> Moreover, the funds are controlled by a multi-stakeholder management board comprised of the North Sulawesi Watersports Association, villagers from the 30 villages in the park, local tourism, fisheries and environmental government agencies, and the local university's marine sciences department. This setup ensures that the money collected is directed to the most important programs needed in the park (as agreed by this diverse set of interests).

Of course, anything described by a promotions board should be viewed with skepticism. An independent party, the Marine Protected Area Management Effectiveness Initiative, vouches that encouraging signs really have begun to develop for the people, reef biota and fishes at Bunaken National Park. The tally of significant conservation accomplishments that have been recently secured in Indonesia includes greater cooperation and heightened collaboration within the advisory group. Park fees are being charged, and the proceeds are put to work to enhance attributes of Bunaken. Local citizens work side by side with water police and park rangers to patrol against theft and resource damage; there has been a near cessation in destructive fishing; and fish abundance and coral cover have both increased throughout the park, with long-term significance for the integrity and health of this potential marine money farm.

So we don't just have to take the word of a promotions board. People with a stake in a positive outcome that is good for both the wild creatures of a park and themselves are putting their hands to work. In the face of what had been a difficult and troubled situation in the early 1990s, not unlike bison being plundered by Hornaday and others on the open range of Montana, an alternative with a great degree of local control and heightened stakeholder responsibility is being established in Indonesia.

To be sure, there are technical, but surmountable, difficulties in defining zones of responsibility and ownership in the oceans. Unlike hard land, which can be enclosed with posts and barbed wire, the soft ocean is difficult to survey and even more difficult to fence. However, the prevailing problems in rebuilding ocean resources lie in government agency intransigence and overcapitalized fishing fleets, not technical concerns.

John Kearney, program leader for the Centre for Community-Based Resource Management at St. Francis Xavier University in Nova Scotia, Canada, has described the difficulty as well as the necessity of rebuilding a sense of responsibility for fishery resources among First Nations people. Those Native Americans who had customarily fished coastal areas were being excluded by fishermen from both near and far with big boats and big debts to service. Kearney concludes that responsible solutions are going to come when local fisher communities, of both aboriginal and European origin, have legitimate seats at the table in deciding the allocation of fish and shellfish and are not just disregarded by the government fishery agencies.

In New Zealand, steps have been taken by government leaders to recover fish stocks and to better benefit those who rely on fish for their livelihood. Maori communities are among those benefiting from assigned rights to marine resources instead of unconstrained, practically uncontrolled exploitation.

While fish in the world ocean have generally not yet been severely damaged except in near-shore settings and at the discharge zones of a number of great rivers, such as the Mississippi's entrance into the Gulf of Mexico, fish and many other creatures in freshwater settings are being damaged despite the aims of national legislation such as the decades-old Clean Water Act of the United States. On the one hand, citizens of the United States and of many other wealthy, democratic nations are paying huge annual sums and making large capital investments for cleaner water confined in pipes. Simultaneously, they are either neglecting or deliberately avoiding the use of meaningful, principled, practical approaches to reduce damage to unconfined, non-point-discharge water. The distinction is between control of point discharges and failure to manage non-point discharges for the public benefit.

Likewise, although on a smaller scale, conservation of waters in particular watersheds may require decades of toil, negotiations and sweat as well as millions of dollars raised by private citizens and public agencies, only to have conservation achievements jeopardized by nonconserving water users. The distinction is between conservation commitment and disregard for both human and biotic community. We have yet to create and test policies that keep rainwater from being damaged as it runs across and through working lands, because we have yet to think in terms of the necessary partnership between those who work on the lands and the rest of the citizenry who would benefit from better watersheds. This arena will be a great challenge to practitioners of hands-on environmentalism in coming decades, as fertilizer use continues at levels beyond crop need and as urban and suburban landscapes grow without incorporation of water-quality conserving practices, such as rain gardens that put storm water into the ground and out of harm's way.

Calling these severe contrasts schizophrenic may seem excessive. But otherwise, how do we appropriately characterize reducing a blue-water river to a sewer for streets and subdivisions, a bank-bound discharge "pipe" for excess anhydrous ammonia and livestock feces? What better label is there for politically correct behavior that clouds the stream with silt and clay and damages the spawning grounds of the watershed fish, said to be held in trust for all the people by the government agency assigned to safeguard the people's water?

We have not yet fully embraced our ownership responsibilities for resources held in common even if they are labeled as "held in trust." Because we sustain illusions of government agency effectiveness in the face of legislative indifference and intransigence, we are willing to be environmental spectators rather than citizens and owners. Hands-on environmentalism is for the purpose of enhancing wildlife, regaining severely depleted fisheries, sustaining watersheds and other interacting communities of humans and biota and, contrary to much of our current practice, integrating the responsibility of ownership with the practice of sympathetic citizenship.

This enlightened citizenship, which creates better health on the land and stronger benefits to those who work the land, is on display across the United States and the entire world. It is active from the Mississippi Delta, where landowners are reclaiming forested wetlands with the help of the scarce Louisiana black bear—a threatened species that is a landowner's asset—westward to the Ponderosa-pine-clad hills of the White Mountain region in Arizona region, where one group of Apache people have created a stunning wildlife management success. There, record-book elk are taken each year with profit to the whole tribe and jobs going to tribal members.

Out of Africa's thin, often barren soil, herds of native wildlife and their predators are being returned to the land because landholders find tangible benefit from the wallets and purses of visiting wildlife viewers who ordinarily live thousands of miles away in the cities of Europe and the Americas. Some of those conserving landholders are families or individuals collaborating actively with neighbors to bring back an array of native wildlife, such as at the Save Valley Conservancy in Zimbabwe. Some landholders making a success of wildlife conservation are villagers in Namibia,

Botswana or Zambia, who are finding ways through the conservancy movement to make local democracy work simultaneously for wildlife enhancement, an increase in family income, and growth in essential community services like schools and clinics. Community decision making will be a vital part of successful hands-on environmentalism.

Dams that blocked migrating fish for more than a century also raised safety and cost issues for a municipality in Wisconsin. In the face of a determined minority who pleaded for an expensive dam replacement program without providing any funds, elected leaders of the city of Baraboo, Wisconsin, took a bold, conservation-minded position. Their decision to partner with various private groups and public agencies to remove unsafe dams has improved river quality, enabled spawning fish species to return, increased the likelihood of meaningful urban rehabilitation in the river edge of the city, and eliminated a significant safety hazard by partnering with private conservation groups and public agencies. The city's dams were replaced at low cost by dam rubble rebuilt into rocky runs. And now that more people can see the moving river, they can focus greater interest on river improvement. The dams that had been assets for the city in its early years became liabilities, and with their removal, the Baraboo River itself is in a much better position to become an enduring, productive community asset.

A collection of electrical power companies led by one firm, Wisconsin Energy, partnering with an internationally active nonprofit, the Nature Conservancy, initiated a large-scale, strictly voluntary enterprise that will remove considerable carbon dioxide from our atmosphere and hold it for a long time in the vegetation of a recovering rainforest preserve and sustainable-use forest in Belize. This is evidence that recognition of environmental responsibility and opportunity for improvement doesn't have to be tied strictly to a certain place when mobile resources like air and water need to be better managed.

Hands-on environmentalism has developed these and other examples of leading landowners making their communities better places. Landowners have often worked with their neighbors and fellow citizens to improve watersheds, bring back fisheries, make imperiled species secure, reduce water pollution and, of necessity, make a buck or save a buck that might be spent unnecessarily.

Our conservation pathfinders in this book have amply demonstrated that, in spite of competing demands on limited human resources, it is possible to construct agreeable, voluntary, enduring solutions to pressing environmental problems. The widespread myths of a need for government mandates to control behavior pertinent to environmental quality or of government as the preeminent provider of recreation areas and the like are unfortunate and destructive. We need to sweep them into the dustbin. Given a chance, a choice and encouragement, landowners, water-rights owners, factory owners, power producers and myriad other of our fellow citizens and neighbors will create greater biotic value in the environment we share.

There is another value derived from hands-on environmentalism in action: cost-savings and frugality. As shown in the case of the Baraboo River restoration, low-cost provision of essential ecosystem services and sought-after amenities can best be developed in community settings where those most affected secure the most benefits. Put another way, those most directly responsible might have to pay great costs and thus can be expected to innovate.

A compelling reason for sweeping away the "we need government to do it" thinking is that even if government action could meet our expectations in this realm, we can't afford this way of thinking. It is not just old and tired, it is also unduly expensive. At another time of world crisis, severe economic difficulties and government budget constraints, during the 1930s and 1940s, Aldo Leopold recognized the need to be concerned about undue tax burdens in several of his journal articles and essays.

Although many environmental groups in the developed nations of the world continue to advertise with apocalyptic warnings and calls to enter one more environmental battle, there is overwhelming evidence that many environmental circumstances have improved significantly in the past century. Not just bison but other wildlife have been brought back from the brink. Whether for wildlife enhancement or water-quality improvement, the time is right to move beyond mandates from central authorities as the overarching approach to environmentalism. Using money more effectively and building stronger social commitments to needed environmental improvement are just two of the reasons to expand the support for and application of hands-on environmentalism.

Around the globe at this time, environmental difficulties thought to be intractable are being resolved and instances of "insatiable" resource use are being mitigated at the local level through principles consistent with the land ethic of Aldo Leopold. Through the connection of responsibility with meaningful rights, the use of local information, adaptive management and good science, joined with a mutual commitment to the human community and the biota of their properties, landowners are engaged with fellow citizens and government agencies as partners in meaningful "hands-on environmentalism."

Index